EXPLORE THE WORLD OF BIOLOGY

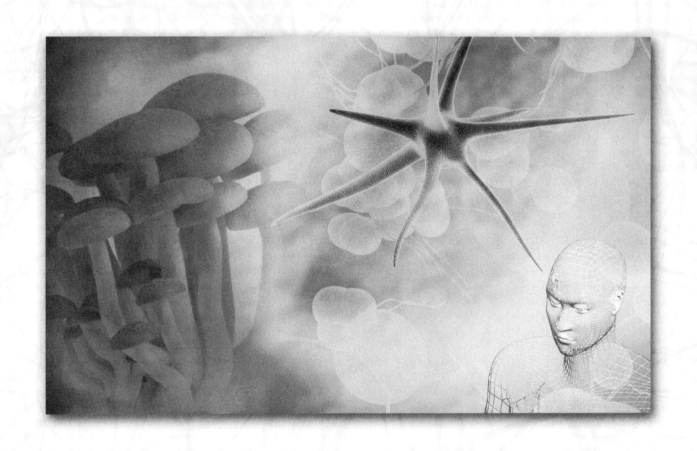

From Mushrooms to Complex Life Forms

JOHN HUDSON TINER

EXPLORING

The World of Biology

First printing: November 2008
Third Printing: October 2013

Copyright © 2008 by John Hudson Tiner. All rights reserved. No part of this book may be used or reproduced in any manner whatsoever without written permission of the publisher except in the case of brief quotations in articles and reviews. For information write:
Master Books®, P.O. Box 726, Green Forest, AR 72638.

Master Books® is a division of the New Leaf Publishing Group, Inc.

ISBN-13: 978-0-89051-552-5
Library of Congress Catalog Number: 2008940597

Interior Design and Layout: Bryan Miller
Cover Design: Diana Bogardus and Terry White

All Scripture is from the New International Version of the Bible, unless otherwise noted.

Please consider requesting that a copy of this volume be purchased by your local library system.

Printed in the United States of America

Please visit our Web site for other great titles:
www.masterbooks.net

For information regarding author interviews, please contact the publicity department at (870) 438-5288.

This book is dedicated to Megan Elizabeth Stephens.

Photo Credits
Shutterstock: 6, 8, 19, 22, 24, 26, 28, 29, 30, 32, 34, 35, 37, 38, 39, 40, 45, 46, 47, 48, 50, 51, 55, 56, 64, 66, 71, 74, 75, 78, 79, 80, 83, 87, 88, 89, 91, 92, 93, 94, 96, 97, 100, 101, 103, 105, 106, 107, 116, 118, 120, 121, 122, 123, 124, 125, 126, 127, 131, 133, 134, 136, 144, 145, 146, 146, 147, 148
I-Stock: 6, 10, 11, 19, 34, 40, 42, 43, 44, 47, 53, 57, 94, 96, 104, 107, 121, 128, 129, 130, 132
Bryan Miller: 8, 20, 23, 25, 34, 43, 59, 61, 62, 65, 68, 75, 77, 86, 102, 117, 138, 139
Science Photo Library: 10, 52
Wikipedia: 17, 114
National: 70
NASA: 58

Table of Contents

How to Use *Exploring the World of Biology*. 4

Biological Classification and Nomenclature. 5

Chapter 1: The Hidden Kingdom. 6

Chapter 2: The Invisible Kingdom 16

Chapter 3: Exploring Biological Names. 28

Chapter 4: Growing a Green World 38

Chapter 5: Food for Energy and Growth 48

Chapter 6: Digestion 56

Chapter 7: Plant Inventors. 66

Chapter 8: Insects 74

Chapter 9: Spiders and Other Arachnids 84

Chapter 10: Life in Water 92

Chapter 11: Reptiles 100

Chapter 12: Birds. 110

Chapter 13: Mammals 120

Chapter 14: Frauds, Hoaxes, and Wishful Thinking. 132

Answers 142

References 149

Index 150

Note to Parents and Teachers:
How to Use *Exploring the World of Biology*

Students of several different ages and skill levels can use *Exploring the World of Biology*. Children in elementary grades can grasp many of the concepts, especially if given parental help. Middle school students can enjoy the book independently and quickly test their understanding and comprehension by the challenge of answering questions at the end of each chapter. Junior high and high school students can revisit the book as a refresher course.

In addition, sections marked "Explore More" can be a springboard for additional study. "Explore More" offers questions, discussion ideas, and research for students to develop a greater understanding of biology.

Biological Classification and Nomenclature

For most of history, biologists used the visible appearance of plants or animals to classify them. They grouped plants or animals with similar-looking features in the same family. Starting in the 1990s, biologists have extracted DNA and RNA from cells as a guide to how plants or animals should be grouped. Like visual structures, DNA (deoxyribonucleic acid) and RNA (ribonucleic acid) reveal the underlying design of creation. Because of the recent switch to DNA and RNA, biological classification is the subject of ongoing debate and proposed changes. Classification is in a state of flux and will remain so for many years. The discussion in this book follows the most settled form of classification, the five-kingdom system, proposed in 1968, which has become a popular standard and is still used by many biologists. The nomenclature uses the English equivalent of Latin terms whenever they are similar.

Chapter 1

The Hidden Kingdom

Classification is the process of grouping objects based on their similarities. For most of history, biologists organized the living world into two kingdoms of plants and animals. Grouping living things into either plant kingdom or the animal kingdom made their work easier. Biologists could easily grasp the broad design of living organisms.

Biologists put into the plant kingdom life that can make food from nonliving material. Chlorophyll (KLOR-uh-fil) is a chemical that gives plants their green color. Plants use chlorophyll and the energy of sunlight to combine water and carbon dioxide to make simple sugars. The process is called photosynthesis (foh-toh-SIN-thuh-siss). Plants use the sugar for growth and to supply energy for building other chemicals, such as cellulose, which makes their cell walls.

Explore

1. What were the first two categories of living things?

2. Why were mushrooms difficult to classify as plants?

3. What classification did scientists give to mushrooms?

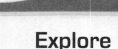

A variety of woodland mushrooms

Biologists put into the animal kingdom forms of life that have sense organs to detect what is around them. Most animals can see and hear. They have a nervous system to interpret what they sense and react to the presence of food or danger. They can move about. Animals cannot make food directly from nonliving minerals. Instead, they must eat plants or other animals.

Biological classification is a system developed by biologists based on their studies and opinions. Once an idea has been accepted for a long time, scientists are reluctant to make changes. From the time of the ancient Greeks — about 400 B.C. — the entire living world was considered made of either plants or animals. But some forms of life, such as mushrooms, did not easily fall into either category.

Although mushrooms looked like plants in some ways, they differ from plants in other ways. The greatest difference was that mushrooms did not have chlorophyll. Mushrooms did not need light. They could live quite well in dark caves, provided they had a source of dead plant or animal matter.

To preserve the two-kingdom classification system, most biologists stubbornly kept mushrooms in the plant kingdom. They insisted upon describing mushrooms as plants without chlorophyll.

The word *mushroom* comes from a French word meaning "moss" or "foam." It is a good choice, because mushrooms are light and airy. Since ancient times, mushrooms have been added to foods to give them a distinctive texture or pleasing taste.

However, scientists did not study mushrooms in detail until the 1700s. They knew that plants had cell walls made of strong cellulose. This gave the cell strength and allowed plants, such as trees, to grow tall despite their immense weight. But study showed mushrooms did not have cellulose.

Biologists discovered that mushroom cell walls were more like those of animals than plants. In addition, mushrooms, like animals, absorbed nutrients from other plants and animals. But biologists could not group mushrooms with animals. Mushrooms had no sense organs, no nervous systems, and no way to move about.

Mushrooms were members of a larger group of similar organisms known as fungi (FUHNG-gye). The singular of fungi is fungus (FUHN-guhss). As biologists learned more about mushrooms, they realized that mushrooms and other fungi did not fit well in either the plant or animal kingdoms.

By the 1960s, biologists agreed that fungi needed a kingdom of their own. They created kingdom Fungi. Into the fungi kingdom they put mushrooms, puffballs, yeasts, molds, mildews, and truffles. The word *fungi* comes from

BASIC STRUCTURE OF A MUSHROOM

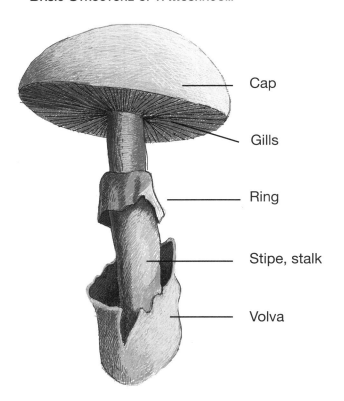

- Cap
- Gills
- Ring
- Stipe, stalk
- Volva

fungi. The underground part of the mushroom expands outward in search of more food. The center of the mat of hyphae dies as food is consumed.

As the hyphae grow, they send up mushrooms that are visible above ground. The mushrooms form a circle along the edge of the expanding mat of living hyphae. In the center, where the food is exhausted, no mushrooms grow, and a ring of mushrooms forms. An average-size ring of mushrooms can be a yard or so across. But with the right growing conditions and abundant food, mushroom rings can be traced out in a giant circle nearly a quarter of a mile across. In a large, grassy field, a single underground mushroom mat can cover as much as 30 acres. The mat of hyphae and the mushrooms it sends to the surface can be one of the largest living things on earth.

a Greek word meaning "sponge." The best-known types of fungi include the yeast that makes bread rise, molds that grow on the surface of fruit and cause it to spoil, and mushrooms.

The stalk and umbrella-shaped cap of a mushroom that we see above ground is only a small part of the entire mushroom. The main body of a mushroom is underground. It is made of a huge network of fine, rootlike threads called hyphae (HIGH-fee). The word *hyphae* is from a Greek word meaning "web." The web, or mat, of hyphae provides food for the mushroom.

Fungi do not make their own food as plants do. Nor do they ingest food as animals do. Instead, hyphae release enzymes (EN-zimes) around dead plant or animal matter. Enzymes are chemicals that hasten chemical reactions. Fungi enzymes break down complex compounds into simpler ones and change them into a liquid. Once nutrients become liquid, they are absorbed through thin cell walls of

This puffball is releasing spores, which are so tiny and numerous they appear to form a cloud.

The above-ground stalk and umbrella — what we usually think of as a mushroom — is a spore case. Spores are one of the ways that mushrooms reproduce. A spore is a tough bit of living matter that can go a long time without water or warmth. Spores grow in the caplike top of mushrooms.

Most mushroom caps have small slits, called gills, on the bottom that release the spores. However, puffball mushrooms have no gills but grow spores inside the ball. The puffball is the champion at releasing spores. Scientists estimate that a single puffball may contain a thousand million of the nearly invisible spores. Millions are released if a person or animal steps on the puffball. Even the pressure of a single raindrop falling on a ripe puffball sends out a cloud of spores.

Spores are lightweight and drift along on air currents. Of the millions released, only a few find the right growing conditions. However, a single spore can grow into a new individual.

Although some mushrooms are good to eat, the actual nutritional value of mushrooms is slight. A bowl of mushrooms has only about 40 calories of food energy. Instead, mushrooms are used to give flavor and texture to other foods.

But most mushrooms do not make good food. They have a bad odor, a tough texture, or disagreeable taste. Others may be all right to eat when fresh, but like other foods, they cause an upset stomach if eaten after they spoil.

A few mushrooms are definitely poisonous. The death cap is the most dangerous. It contains a toxin that causes the stomach lining to dissolve. It destroys the liver and kidneys. About half the people who eat death cap die within 24 hours.

Some people use the name toadstools for poisonous mushrooms. However, this has no scientific basis. Mushrooms and toadstools have no physical differences. Toadstools have no visible feature that warns a person they are poisonous. Unless a person can identify a mushroom exactly, the only mushrooms safe to eat are those grown commercially.

Mushrooms for sale in grocery stores are grown in caves, dark cellars, or special barns. The mushrooms grow in beds of rotted manure and straw covered by a layer of soil. Eatable mushrooms are harvested and taken to market while still fresh.

Some types of mushrooms and edible fungi are difficult to grow commercially. Truffles, for instance, are the most highly prized of the edible fungi. But they are so difficult to grow that most are harvested in the wild. They grow underground. They have a distinctive odor. Some people can smell where they are growing beneath the soil. But most truffle hunters use dogs or pigs that have been trained to sniff them out. Pigs not only find the truffles but can also dig them with their strong snouts.

Even if mushrooms are not used for food, they have been given an important role in nature. They help break down dead trees and other plant life and return nutrients to the soil.

Yeasts are fungi too, but they are made of a single cell. Unlike mushrooms, they are invisible to the unaided eye and can only be seen with a microscope. For almost 200 years after the invention of the microscope, most scientists ignored single-celled life. They thought microscopic life could not affect larger forms of life. Louis Pasteur proved otherwise.

Louis Pasteur, the French scientist, was the first person to understand the true nature of yeast. Louis Pasteur is well known today for his many important scientific discoveries. But in 1854 he was a struggling young science teacher.

A businessman came to Louis with a problem. "I produce alcohol from beet root sugar," the man said. "Something is going wrong. The alcohol turns sour and useless. Will you investigate the matter?"

Louis agreed to help the businessman. He learned all he could about fermentation (fur-ment-tay-shun), the process that changes beet sugar to alcohol. He knew that fermentation was an important process. People have used the natural process of fermentation since ancient times. One type of fermentation changed milk to cheese. Another type changed flour into light and tasty bread. Fermentation could also be a nuisance by changing fresh milk into sour milk and sweet grape juice into sour vinegar.

People used brewer's yeast to make alcohol and baker's yeast to change flour to bread. Yet no one understood how yeast worked. Chemists believed yeast to be nothing more than a complex chemical substance. Most said, "Fermentation is purely a chemical process." Chemists wrote this in textbooks and taught it in classrooms.

Truffles are edible fungi that grow underground.

Louis Pasteur collected samples of beet juice at various stages of its manufacture: at its start, when well underway, and at its completion. He took samples of the good juice and of the bad. He peered for hours at samples of beet juice through a simple student's microscope.

Almost at once he saw ball-like yeast cells in the liquid from the good samples. He noticed that a drop of beet juice swarmed with yeast cells. As he watched, a yeast cell sprouted a bud that grew larger. Then it broke away from the parent cell. Pasteur saw that yeast grew and reproduced by budding. A small bud formed on the surface of a parent cell. The daughter cell enlarged, matured, and detached from the parent.

Reproduction from a single parent is asexual reproduction. In asexual reproduction, genetic material comes from only

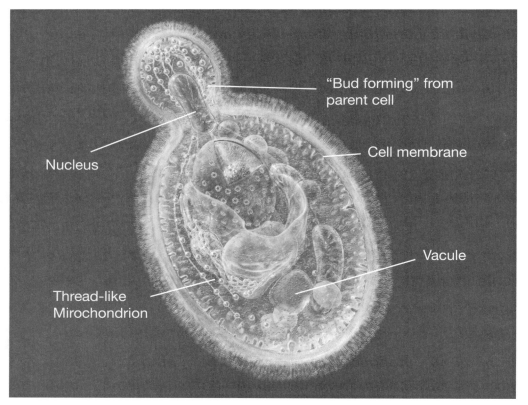

Yeast grow and reproduce by budding.

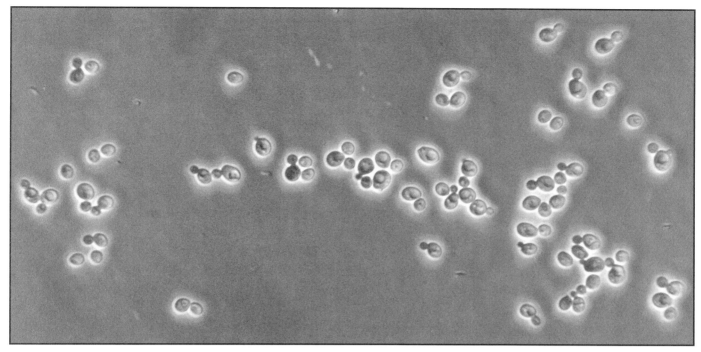
Photomicrograph of the yeast form *Candida albicans*, a disease-causing fungus.

one parent. The daughter cell of yeast is identical to the parent cell.

"The yeast cells are alive," Louis whispered. "They grow. They multiply."

Pasteur proved that yeast cells were living things. Until Louis investigated, scientists did not know that yeast was alive, nor did they know that tiny living cells could change sugar into alcohol. Louis showed that yeast cells consume sugar as food and change it into carbon dioxide and alcohol.

Fermentation of beet juice produced carbon dioxide gas and alcohol. Louis could tell how well fermentation was going by how much carbon dioxide gas bubbled to the surface. Fermentation took place most rapidly while the beet juice teemed with ball-like yeast cells. Fermentation stopped when yeast died.

He compared the good juice with the bad. In juice from the spoiled vats, he found only a few ball-like yeast cells. Instead, a vast number of rod-shaped cells grew all through the sour beet juice. Now he knew the reason for spoiled juice. Two types of microscopic life lived in the juice and competed with one another. If the yeast won, then the juice remained good. If rodlike microorganisms prevailed, they changed sugar and even alcohol into lactic acid, which is sour.

He explained his discovery to the businessman. "It's like a war between two types of microscopic organisms that grow in the beet juice. Round cells change sugar into alcohol. When rodlike microorganisms take control, they produce lactic acid instead of alcohol."

Louis Pasteur showed that as yeast received their nutrition, they changed complex chemicals into simpler substances. For instance, yeast changed the carbohydrates of flour into alcohol and carbon dioxide. Yeast caused bread to rise by releasing carbon dioxide gas. The gas produced

Pasteurization

One morning as Louis Pasteur's wife fixed breakfast, she drew back in distaste as she sniffed the milk. "The milk has spoiled."

"Why does milk sour?" Louis wondered. He had found that microscopic life forms could cause beet sugar to become sour. "Could tiny microorganisms be the cause in milk, too?"

Louis studied how the milk soured. As he suspected, he found tiny microorganisms that flourished when the milk turned bad. He could see them budding and multiplying.

Louis told his wife, "I have found the cause, but now I must learn how to prevent the milk from spoiling."

Louis knew that microorganisms grew when the conditions were right for them. After a series of experiments, he learned that gentle heating of the milk destroyed the tiny life. Yet the milk still stayed fresh and tasted the same.

The process became know as pasteurization. It could be used to preserve countless perishable beverages and foods, including milk, cider, and cheese. Pasteurization became a household word. Louis Pasteur became famous around the world.

"Pasteurization will make you a wealthy man," his friends assured him. As a schoolteacher and part-time scientist, he had a meager income. Would this discovery be the end of his financial worries?

"No," Louis decided. "I became a scientist to bring science to the benefit of mankind. I will give the discovery to the public."

Louis based his decision upon the teachings of Jesus. He believed he had a duty to help those in need. Pasteurization turned out to be an important advance in the history of science. Yet Louis Pasteur made not a cent from the discovery.

Some scientists resented his success and called it mere luck. Louis Pasteur replied promptly, "Luck favors the prepared mind."

One way he prepared his mind was with Bible study and prayer. Louis often spoke to his family about the importance of Christian faith. In letters to his sisters, he told how he read the New Testament to let its simple truths guide his life. He encouraged them to pray for one another.

Despite his long hours of research, Louis continued to teach and train young people. He told his students, "I say to each of you, do not let yourself be tainted by that spirit of disbelief which likes to belittle everything.

"Say to yourself first of all: 'What have I done to educate myself?' Then, as you advance further, ask yourself, 'What have I done for my country?'

"But whatever happens, whether your work succeeds or fails in the test of life, the most important thing of all, as one approaches the end, is to be able to say, 'I have done what I could.' "

Louis' study of science helped him admire and worship God. Louis said, "The more I study nature, the more I stand amazed at the work of the Creator. Into His tiniest creatures, God has placed extraordinary properties."

tiny pockets in the dough. The alcohol in bread was driven off as it was baked. The alcohol gave baking bread a pleasant aroma.

Louis Pasteur was the first scientist to call attention to the power of microscopic creatures. He proved that germs could cause infection and disease in animals and even in humans. He said, "The role of the infinitely small in nature is infinitely great."

Molds, like mushrooms and yeast, are also members of the fungi kingdom. Although small, molds are just visible to the unaided eye. They grow on bread that is moist and warm. Bread molds viewed through a magnifying glass appear similar to miniature mushrooms. Molds appear as small balls on tiny stalks. They have threadlike structures that cover the surface of the food and extend below the surface.

In humid climates, molds grow on any surface that is made of something that was once alive; that is, an organic (or-GAN-ik) substance. An organic substance is one that comes from a living source. Molds grow on clothes made of cotton and on shoes and purses made of leather. They also grow on food, especially fruits that have high sugar content. As molds digest sugar, fruit becomes soft and mushy.

Molds, like mushrooms, reproduce by spores. Molds have a knoblike spore case at the top of a stalk. After the knob becomes ripe, it breaks open and releases spores. Airborne spores fall upon organic material and begin growing. If spores fall on cotton, leather, or anything else from a plant or animal, they can begin growing, especially in a warm and humid climate.

Molds cause bread to spoil. Mold needs moisture to thrive. In the past, bread was toasted to make it crusty all the way through. Even today, the hard, dry crust of bread slows mold from taking hold.

Although not always poisonous, some mold makes food taste bad. On the other hand, humans eat some types of mold as food. Roquefort cheese contains a blue-green mold that gives it a sharp flavor.

Penicillium (pen-uh-SIL-e-uhm) mold is another example of a useful mold. It produces penicil-

Mold on cheese

Alexander Fleming

Alexander Fleming, a British medical researcher, discovered penicillin. He worked at St. Mary's Hospital in London, England. During the warm days of September 1928, Fleming raised the laboratory windows.

Alexander Fleming experimented with Staphylococcus (staff-i-low-COCK-us) bacteria that cause boils and other infections in humans. Staph infections could be deadly. He grew the bacteria in circular, flat dishes designed to hold a jellylike food for growing bacteria.

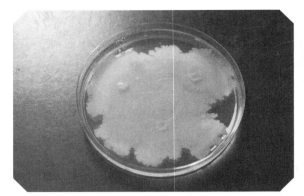
Petri dish with bacteria colony growing in it.

One day he noticed a bluish mold growing in one of the dishes. A mold spore had probably floated in through the open window from outside and begun growing as opportunity provided.

Fleming examined the border of the mold. *That's funny*, he thought. *Yellow colonies of bacteria should be growing all around the mold*. Instead, the staph bacteria had been dissolved by the mold or something the mold released.

He transferred a bit of the mold to nutritive soup. Within days, the mold covered the surface of the Petri dish. Fleming drew off a bright yellow liquid released by the mold. He found that the golden liquid alone would kill staph bacteria. Fleming diluted it a thousand times in water. Even in the dilute form, it proved to be a powerful germ killer. It killed not only Staphylococcus, but also several other types of bacteria.

Apparently in its competition with bacteria for the same food, molds release chemicals that kill bacteria and eliminate the competition.

Fleming gave the name penicillin (PEN-ah-sil-in) to the active agent released by the mold. Tests showed that the penicillin produced by the mold would kill germs but not damage human cells. Penicillin came into widespread use during World War II. Since then, several other molds have been found with the ability to destroy bacteria that cause diseases.

Fleming's keen observation overcame primitive conditions in a poorly equipped laboratory to make the most important disease-fighting discovery of the last 100 years.

lin, a powerful treatment to fight bacteria that cause infections. Because penicillin kills bacteria, it is called an antibiotic drug. The word *antibiotic* comes from *anti*, meaning "against," and *bio*, meaning "life." In this case, the life that penicillin acts against is small bacteria.

Bacteria are tiny, one-celled creatures even smaller than fungi. Bacteria are not members of the fungi kingdom, but many of them do compete with fungi for the same food.

Discovery

1. Early scientists classified all livings things as plants or animals.

2. Mushrooms do not make food by photosynthesis.

3. Mushrooms are members of the Fungi kingdom.

Questions

T F   1. For most of history, living things were classified as either plants or animals.

A B   2. Mushrooms were studied in detail by (A. the Greeks in 400 B.C.   B. scientists in the 1700s).

A B C D   3. To keep mushrooms in the plant kingdom, scientists described mushrooms as plants without (A. cell walls   B. chlorophyll   C. seeds   D. sunlight).

A B C D   4. Today, biologists classify mushrooms as members of the (A. animal kingdom   B. bacteria kingdom   C. fungi kingdom   D. plant kingdom).

T F   5. The only way mushrooms can reproduce is by sending out hyphae.

T F   6. The mat of hyphae and the mushrooms it sends to the surface can be one of the largest living things on earth.

A B C D   7. The above-ground stalk and umbrella of a mushroom is used to (A. absorb carbon dioxide   B. catch sunlight   C. release spores   D. sense the presence of enemies).

A B   8. Pigs are used to hunt for (A. truffles   B. death cap mushrooms).

A B   9. Louis Pasteur realized yeast cells were alive when he saw them (A. cause milk to sour   B. grow and reproduce).

A B C D   10. What do yeast cells consume as food? (A. alcohol   B. carbon dioxide   C. sugar   D. vinegar).

A B C D   11. Fungi that are growing on bread and are just visible to the unaided eye and look like miniature mushrooms are most likely: (A. lichen   B. mold   C. truffle   D. yeast)

A B   12. The mold that grew in Alexander Fleming's dish was there because (A. he was experimenting with bread mold   B. it probably drifted in through an open window.)

Explore More:

Explore More is an opportunity to explore the subject in your own way. Take a photograph, draw a picture, collect a sample, make a poster, write a poem about the subject, list the pros or cons as to whether the subject is helpful or harmful, or interview a person who has experience with the subject. For example, interview a person who has had pneumonia. How did the doctors treat the disease? Have you ever eaten Roquefort cheese? How would you describe its taste?

Subjects for more exploration:

lichen, rust (plant disease), mildew, Dutch elm disease, Roquefort cheese, pneumonia, penicillin-resistant diseases

Chapter 2

The Invisible Kingdom

Although mushrooms were known from ancient times, another form of life was not discovered until the 1600s. Until the invention of the microscope, scientists did not suspect that living things could be so small as to be invisible to the unaided eyes. The first scientists to see the tiny life were Anton van Leeuwenhoek (LAY-ven-hook) of Holland and Robert Hooke of England.

Robert Hooke served as chief experimenter for the Royal Society. Scientists in this organization met in London to discuss the latest discoveries in science. The Royal Society had the motto of "Nothing by Mere Authority." Even the most renowned scientist had his new discoveries tested to ensure their accuracy.

Explore

1. What invention made little life visible?

2. Can life invisible to un aided eyes be the cause of disease?

3. Why are protista and bacteria different from plants and animals?

Robert Hooke and His Microscope

Although Anton Leeuwenhoek made amazing discoveries with his simple lens microscopes, Robert Hooke's design was clearly better. It is the one used today. It has two sets of lenses. One is small but powerful, and placed near the subject being viewed. This lens makes an image that is viewed by a second group of lenses next to the eye.

With his improved microscope, Robert Hooke looked at insect wings, feathers, fish scales, and tiny living things, such as molds and mosses. He was an exceptionally good artist and sketched what he saw in vivid detail. This was before the invention of photography.

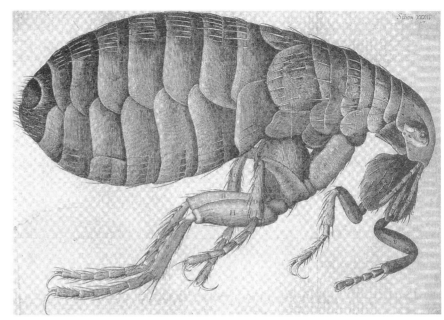

Drawing of a flea done by Robert Hooke and published in his book *Micrographia* in 1665. Hooke was curator of experiments at the Royal Microscopical Society of London, which was founded in 1660. He used a compound microscope of his own design to view a wide range of specimens. He then drew and published the results of his remarkably detailed observations.

His drawings brought before the public the incredible detail found in the smallest living things.

In 1665 Robert Hooke published a book, *Micrographia*, about his tiny subjects. His portrait of a flea is often reproduced today. Naturalists hailed Robert's book as a major achievement in understanding biology.

Robert Hooke also invented the term *cell* for the basic building blocks of life. He cut a thin slice of cork and examined it under his microscope. He noticed a regular pattern of honeycomb-like pores. They reminded him of a series of tiny empty rooms. He called them cells.

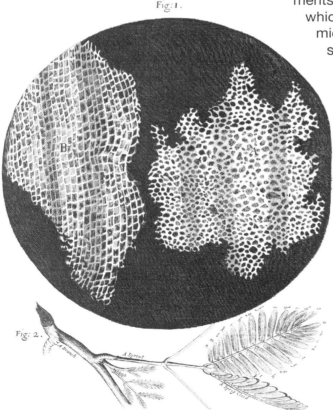

Drawing by Robert Hooke of cork seen under a microscope; the first picture showing "cells" in a biological specimen, named by Robert Hooke.

The Royal Society employed Robert Hooke to repeat experiments and prove or disprove the claims of fellow scientists.

The Royal Society received letters from Anton van Leeuwenhoek, an amateur scientist living in Holland. Leeuwenhoek's hobby was making microscopes and viewing small things of nature through them. He examined wings and eyes of insects, scales and fins of fish, and many other subjects.

Leeuwenhoek sent letters describing what he saw to the Royal Society. The Royal Society, in turn, asked Robert Hooke to repeat his observations. Hooke agreed that the amateur scientist's work was of the highest standards.

In 1674 Leeuwenhoek wrote about an astonishing discovery. He claimed to have found a world of living things inside a drop of canal water. Although invisible to the unaided eyes, they moved about, grew, ate, and reproduced. They produced offspring by splitting. When the temperature and food were right, they grew and split quickly. He calculated that a single drop of water would be home to one million of the little animals. He called them animalcules.

Robert Hooke tried to see the little life, but his microscope did not magnify as much or produce as sharp an image as those made by the Dutchman. Some members of the Royal Society expressed doubt that the little life existed. Robert Hooke asked Anton Leeuwenhoek to send one of his microscopes to the Royal Society, which he did.

With the borrowed microscope, Robert Hooke managed to verify most of Leeuwenhoek's claims. But using the tiny lenses required patience and keen eyesight. Robert designed an improved microscope. Instead of a single powerful lens magnifying 100 times, he combined two weaker lenses to give the same magnification. He also arranged mirrors to focus light from a window on the subject being magnified and make the image brighter.

With the improved microscope, he saw Leeuwenhoek's tiny animalcules. He found them in water in rain barrels, ponds, and canals. They were not all of the same size or shape.

Meanwhile, in Holland, Leeuwenhoek made yet another discovery. His best microscope magnified 270 times. At the very limit of its power, and his skill at using it, Leeuwenhoek detected another category of small life. He saw rodlike, ball-like, and corkscrew-shaped objects far smaller than anything he had seen before. Yet they lived, grew, and moved about on their own, too.

Even Robert Hooke failed to detect the even

Student using a compound microscope

smaller life. A hundred years passed before microscopes became powerful enough to confirm what Leeuwenhoek had seen. Leeuwenhoek had become the first to detect bacteria. Another 100 years passed before scientists realized that despite their small size, bacteria could cause disease.

Most of the small, single-celled life visible through the microscope appeared to be animal-like. They could move about and had no chlorophyll. Biologists gave the name protozoa (proh-tuh-ZOH-uh) to Leeuwenhoek's little animalcules. The *zoa* part of protozoa means "animal." It is the same root word that gives us the word *zoo*, a place where animals are kept. The singular of protozoa is *protozoan*. At first, scientists thought protozoa should be classified with animals, but this idea later changed.

A paramecium (par-uh-MEE-shee-uhm) is a protozoa that is shaped like a slipper. The name is from a Greek word meaning "oblong." It can move, capture and digest food, eliminate solid and liquid wastes, protect itself against enemies, and reproduce.

The paramecium has fine, hairlike filaments called cilia that wave back and forth to propel it

Simple and Complex

Occasionally, people will refer to single-celled life as primitive. They may describe paramecia, amoebas, or other micro-organisms as lower forms of life. Terms such as simple and complex, primitive and advanced, and lower forms of life and higher forms of life can be misleading. The phrases have one meaning in daily life but have quite a different meaning in biology.

In biology, a complex organism is one that carries out its life processes by many different systems. Special organs are dedicated to each function. The human body is an example of a complex organism. We have tissues, organs, and systems dedicated to specific functions. Muscles and other special tissues combine to make the heart, which is an organ. The heart and other organs, such as blood vessels, combine to make the circulatory system. Other systems include the nervous system, respiratory system, digestive system, and reproductive system. Each system has a particular job. Our cells are different, too. Skin, muscle, and nerve cells differ from one another. The human body is a complex organism.

Protozoa carry out all of these life functions, too. But they do so within a single cell. Single-celled life, such as a paramecium or an amoeba, does a multitude of tasks without benefit of different organs. Protozoa are described as simple forms of life. However, microorganisms such as protozoa are amazingly complex. They accomplish in a single cell what other living things use millions of cells to do.

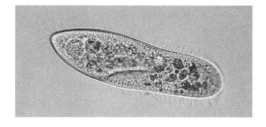

Simple (paramecium)

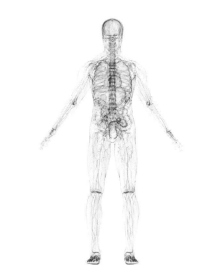

Complex (human being)

through the water. Cilia also sweep food particles into an opening called the gullet. Cilia line the gullet and push food into a vacuole (VAK-yoo-OIL), a pouch-like miniature stomach, to be digested. Digested food is absorbed directly into protoplasm, the living substance of the cell. Because the paramecium lives in water, it must constantly collect and rid itself of excess moisture. The liquid water is collected in vacuoles that can contract to squirt the water out of the paramecium.

The paramecium can detect enemies. It repels them by shooting out tiny bristles like miniature darts.

The paramecium, like other protozoa, has a nucleus. The nucleus controls the activity of the cell, including reproduction. Most protozoa reproduce asexually by fission. Fission means to split apart. A fully grown organism divides into two daughter cells. Each daughter cell has a complete set of genetic instructions for growth and reproduction.

Amoebas (uh-MEE-buhs) are another common type of protozoa. They live in soil and water. The larger amoebas are about one-fortieth of an inch long. When in a drop of water that is illuminated from the side, they are just visible to the unaided eye.

An amoeba has no particular shape but changes form and appears to ooze about. A network of protein fibers builds its structure. As the amoeba changes shape, it takes apart its internal structure

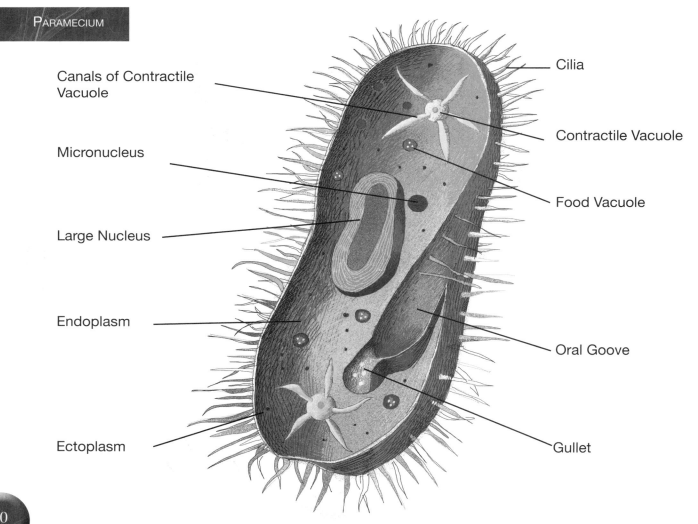

PARAMECIUM

and rebuilds it in the direction it wants to move.

When the amoeba encounters food, it flows around the food particle and forms a vacuole. It pumps enzymes into the vacuole to digest the food. After digestion, it leaves the waste products behind as it moves along.

Protozoa do cause disease, although in the United States most protozoa diseases can be prevented or have effective treatments. In the United States, one well-known disease is giardiasis (gee-are-DIE-sis), caused by the *Giardia lamblia* protozoa. It was one of the first protozoa identified, being described by Leeuwenhoek himself in 1681. *Giardia lamblia* protozoa are parasites that live in the intestines of animals such as cows, deer, and dogs. As a parasite, it derives all of its nourishment from the host and provides no benefit in return.

Giardiasis protozoa survive outside their host as cysts, which contaminate food or water. Hardy cysts can survive outside a host three months or more waiting for another host. Once ingested, they spring back into active life. Hikers and backpackers who drink from streams must treat the water — by boiling or adding a chemical such as iodine — to avoid giardiasis. Otherwise, they will suffer stomach cramps, diarrhea, and nausea.

Worldwide, malaria is the single most deadly protozoa disease. Malaria is usually found in the tropics. The victim shakes with chills, followed by intense sweating from fever. Sometimes the symptoms go away, but they will return. A particular type of parasitic protozoa causes malaria.

As with most diseases, the best solution is not to get the disease in the first place. Because malaria is transmitted by mosquitoes, people in tropic countries sleep under mosquito nets, spray insect repellant, and eliminate pools of water where mosquitoes breed. Malaria has been studied for centuries, but no vaccine exists. Should a person get the disease, he or she cannot be cured, although the chills and fevers can be reduced. Quinine was the first treatment. Quinine is a chemical from the bark of the cinchona tree. Other drugs have been made since then, but they merely make fevers and chills less severe and do not eliminate the disease.

In places where malaria is common, people with the disease miss work, victims must be cared for by others, and the economy suffers. People become poor in struggling against the disease, and that poverty results in unsanitary conditions that lead to even more disease. Finding an effective way to cure malaria is one of the great unsolved problems of biology.

Although paramecia, amoebas, and other protozoa seemed to fit in the animal kingdom, other one-celled life did not. The euglena (you-GLEEN-uh) was the most bothersome one from a classification standpoint. Euglena is found in fresh water. It is a single, microscopic cell that contains chlorophyll and looks green. It makes food by photosynthesis.

But the euglena has a whiplike structure called a flagellum. By rapidly thrashing the flagellum back and forth, the euglena moves through the water. It can move about as other protozoa.

Should the euglena be classified as a plant or animal? Biologists could not agree. Because it could move, some called it an animal. Because it had chlorophyll, others called it a plant.

Algae (AL-jee) are another type of small living things. They contain chlorophyll for photosynthesis. Some form long strands of

Lichen and Microbiotic Crust

Some plants and animals survive in harsh climates by symbiosis (sim-bee-OH-siss). Symbiosis is a close association of two or more living things. Symbiosis means living together. Another name for symbiosis is mutualism, meaning "to borrow from one another." They borrow from one another to the benefit of both.

Some living things are tightly bound together. Lichen grows almost anywhere, even on poor soil and rocks. Lichen looks like a tough moss, but it is a living partnership of fungi and algae. Lichen has a layer of algae sandwiched between two layers of fungi.

Algae contain chlorophyll, the light-absorbing material that makes food from sunlight. Algae take water, minerals, carbon dioxide, and sunlight and makes food that it shares with fungi. Fungi lack chlorophyll. But they release an acid that breaks rocks into simple minerals that algae can use. Fungi also furnish algae with water. In desert climates, the layer of fungi shades algae from the relentless rays of the sun. In cold polar tundra, the layer provides a small measure of warmth. Together, the algae and fungi of lichen survive and even thrive in harsh conditions.

Lichen is effective in weathering that breaks down larger stones into finer grains of minerals. The action of lichen allows other plants to take root and begin growing.

Microbiotic crust is another unusual plant community even more complicated than lichen. It looks like a brown layer on top of desert sand. Microbiotic crust is a living soil. Tiny mosses, lichens, and independently living fungi and algae grow together and help one another.

Microbiotic soil is important to life in the desert. It has sticky roots that attach to sand particles. The hardy covering protects against the ceaseless action of wind. Microbiotic crust holds the soil in place.

Its rootlike fibers serve as miniature water tanks, too. A desert receives less than ten inches of rain a year. The moisture usually comes as a sudden cloudburst. The precious water would quickly drain from hard-packed desert surface. But fungi in microbiotic life take in rainwater.

The living soil makes life possible in areas that otherwise might remain barren. A seed falling on the moist surface is more likely to grow than a seed that falls on parched desert sand. Microbiotic soil provides a small amount of moisture. First, plants such as piñon pine, juniper, sagebrush, and Indian paintbrush take root. Next, animals such as mule deer, kit fox, rabbits, kangaroo rats, and small reptiles live where the plants grow. Flocks of blue jays and blackbirds are attracted to the seeds. Red-tailed hawks hunt rodents, reptiles, and rabbits. A desolate desert becomes a living landscape.

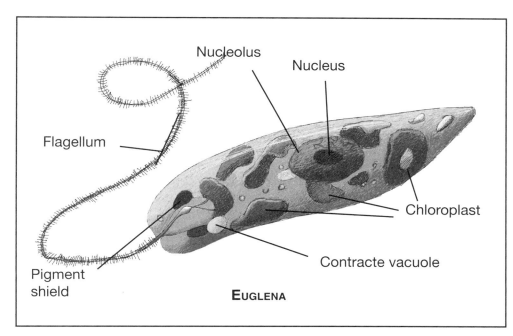

similar cells. But many live independently as a single cell.

Diatoms are single-celled algae that surround their cell wall with a coating of silicon dioxide, the same substance that makes sand. The cell wall is hard as glass but transparent to let in light for photosynthesis. When diatoms die, their microscopic, glasslike, hollow shells are left behind. In some places, remains of diatoms have piled up in deep layers.

Diatom shells are mined and sold as filtering material. Some swimming pool filters use diatoms to remove small, suspended particles and clean the pool water. Diatom cell walls are also used to make sticks of dynamite. Diatom shells absorb oily nitroglycerin to give a paste that is formed into explosive sticks.

Some algae live on land. Moss is a type of algae that grows on rocks and trees. Direct sunlight causes it to dry out and die. Moss lives best on the moist, shaded northern side of trees. Moss on trees can be used as a direction finder. The side of trees where moss grows best faces north.

Algae have a major role in restoring oxygen to the atmosphere. Oxygen is an active element and quickly forms compounds with other chemicals. As animals breathe, they remove oxygen from the air and combine it with other elements. During photosynthesis, plants and algae break down oxygen compounds and restore free oxygen into the atmosphere. Algae, mostly those in the ocean, account for at least half of all oxygen returned to the atmosphere.

Because of the large size of some algae and the presence of chlorophyll, it would seem that algae should be classified with plants. At first, biologists did just that.

But placing them with plants was not entirely satisfactory. Algae have no roots, stems, or leaves. They do not have a vascular system to

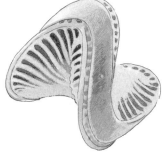

Diatom shells

transport sap. And they can survive as a single cell.

Biologists found three types of small life that could live as a single cell. Some were small, animal-like creatures capable of propelling themselves, the protozoa. Others were plantlike life with chlorophyll, such as euglena and algae. They also discovered some organisms, such as slime molds, that had properties similar to those of fungi. However, none of the small life fit well into plant, animal, or fungi kingdoms.

Algae in a pond

Biologists decided to create still another kingdom, kingdom Protista. They put the protozoa, euglena, algae, slime molds, and other one-celled life into kingdom Protista. Biologists made a rule to distinguish plants and animals from small life. Plants and animals must be multicellular, which means they must be made of more than one cell. A member of kingdom Protista must be made of a single cell.

Biologists now had four kingdoms: Plant, Animal, Fungi, and Protista. But the kingdom making was not yet finished. Biologists had to deal with bacteria. Bacteria come in three basic shapes: spherical, rodlike, and spiral-like. Like some protozoa, they have flagellum for moving about.

Bacteria, in addition to being far smaller than protozoa, lacked some features found in other forms of life. Members of all other kingdoms each have cells with a separate nucleus. The nucleus stores genetic information and copies it at the time of cell division. Although bacteria cells carried genetic material, it was not collected in the central location of a nucleus.

Bacteria have a simple design and small size. Their small size makes it possible for them to grow and divide rapidly. They flourish in almost any type of environment. They are found nearly anywhere in large numbers. They can use practically any organic material for food. Although small, bacteria can move about; they have one or more hairlike flagella to propel them. They can advance toward food and retreat from areas that make them uncomfortable.

Bacteria are found anywhere other things live and in some places where other life would quickly die. Some bacteria do not need oxygen. About half of all bacteria are anaerobic (ANN-uh-ROW-bik), meaning they can live without oxygen from air or water. The word *anaerobic* is made of three parts: an-aero-bic; *an* means "not," *aero* means "air," and *bic* means "life." The word *anaerobic* means "life without air." The bacteria in a septic system that breaks down waste products are of the anaerobic type. In fact, oxygen is a poison to them and will kill them.

Bacteria can cause diseases in humans, animals, or plants. Some of the most deadly diseases of all time, such as Black Death (plague) and tuberculosis (TB), are caused by bacteria.

Other bacteria are especially important to promote growth of certain crops. Plants need several elements to grow. Nitrogen is one of the required elements. Nitrogen itself is plentiful. The earth's atmosphere has four times as much nitrogen as oxygen. But plants cannot retrieve this vast storehouse of nitrogen. Most plants cannot take it directly from the air. Instead, they extract nitrogen from compounds in the soil.

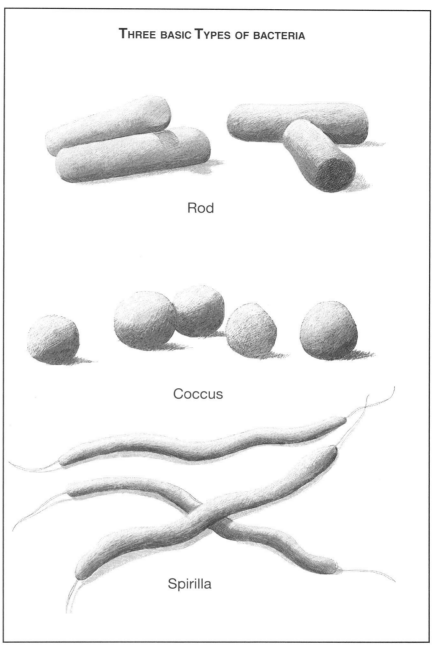

THREE BASIC TYPES OF BACTERIA

Rod

Coccus

Spirilla

As crops are harvested year after year, nitrogen compounds in the soil are removed. But nitrogen-fixing bacteria can replace the nitrogen. One species of bacteria forms colonies on the roots of legumes such as clover, peanuts, and peas. Planting these crops restores nitrogen to the soil.

Bacteria in our bodies also provide us with needed nutrients, such as vitamin K, which the body itself cannot make.

Bacteria serve other functions, too. Like fungi, bacteria feed on dead plants and animals. Without them, plants and animals would not return quickly to the minerals that make them. Instead, they would merely wear away in the same way that wind and water wear away rock. But bacteria hasten the decay of dead organic material and quickly return nutrients to the soil.

Some bacteria, however, have only a glancing impact on our lives. Biologists study bacteria that cause disease or help mankind. The vast majority of bacteria are neither directly harmful nor helpful. Their role in the environment is not as well studied and poorly understood. This is another area of biology that needs to be explored.

Because of their many differences from protozoa and other life, biologists have given bacteria a kingdom of their own: kingdom Bacteria. This kingdom is made of tiny, single-celled organisms whose genetic material is loose in the cell and not in a central nucleus.

Biology classification is not entirely settled. Viruses are 20 to 100 times smaller than bacteria. They cannot be seen in ordinary optical microscopes. Electron microscopes are required to reveal their structure.

They appear to be merely a nonliving ribbon of genetic material surrounded by a protective coating of protein. They cannot grow or reproduce on their own. Once they invade a living cell, they appear to be alive. They take over the cell and use it to reproduce. Because they differ so greatly from other living things, viruses have not yet been included in the classification with any of the other kingdoms or with a kingdom of their own.

Currently, biologists group living things into five kingdoms: plant, animal, fungi, protista, and bacteria.

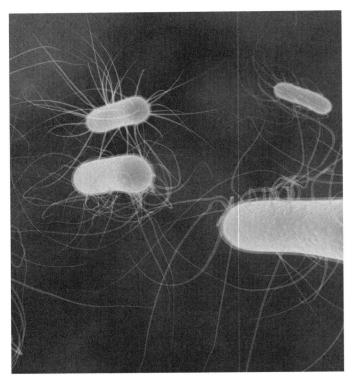

E. coli bacteria is normally present in human intestines, but one particular type can cause severe illness.

Discovery

1. The microscope revealed a world of living life in a drop of water.

2. Protista and bacteria can cause diseases such as malaria and plague.

3. Protista and bacteria are single-celled life, while plants and animals are multicellular.

Questions

A B 1. The Royal Society employed (A. Robert Hooke   B. Anton van Leeuwenhoek) to test claims of fellow scientists.

A B 2. The first scientist to see the little life in a drop of canal water was (A. Robert Hooke   B. Anton van Leeuwenhoek).

T F 3. The paramecium and amoeba were given the name *protozoa* because they appeared to be animal-like.

A B C D 4. The one that can change its shape is the; (A. amoeba   B. euglena   C. giardiasis   D. paramecium).

T F 5. A protozoa is called simple because it cannot carry out all of life's functions.

6. The single most deadly protozoa disease is _____.

A B 7. At first, a euglena was called a plant because it (A. could not move   B. had chlorophyll.)

A B C D 8. Single-celled algae that surround their cell wall with a coating of silicon dioxide are: (A. anaerobic bacteria   B. diatoms   C. giardiasis cysts   D. macrobiotic crust).

T F 9. A member of kingdom Protista must be capable of surviving as a single cell.

10. Lichen is a layer of algae sandwiched between two layers of _____.

T F 11. All living things must have oxygen to survive.

A B 12. Plants need nitrogen to grow, which they must get from (A. the air   B. nitrogen compounds in the soil.)

13. Matching:

a. animal   ____ nonliving genetic material that only comes alive inside a living cell
b. bacteria   ____ multicellular life that can move and has sense organs
c. fungi   ____ multicellular life that includes mushrooms
d. plant   ____ multicellular life that makes food by photosynthesis
e. protista   ____ single-celled life that includes paramecium, amoeba, and euglena
f. virus   ____ single-celled life without a nucleus; one form causes Black Death (plague)

Explore More:

Explore More is an opportunity to explore the subject in your own way. View little life through a microscope. Describe what you see. Research the prevention of diseases caused by protozoa, bacteria, and viruses. What are the risks and benefits of protista and bacteria? Read about Robert Hooke and Louis Pasteur. What important discoveries did they make? How does vaccination prevent a disease? What is the difference between the prevention of a disease and the treatment of a disease? How do outdoor experts recommend treating drinking water when backpacking?

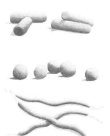

Chapter 3

Exploring Biological Names

Biologists who study plants and animals give them names. The Bible says that after forming the beasts of the field and fowls of the air, God brought them to the man (Adam) to see what he would name them. "Whatever the man called each living creature, that was its name" (Gen. 2:19).

Ever since then, people have been giving names to plants and animals. It is interesting to play detective and trace down how names have been given to animals. For instance, the Canary Islands are a lonely group of islands

Explore

1. Are canary birds named after dogs?

2. Is a panther a particular breed of cat?

3. What is the goal of biological classification?

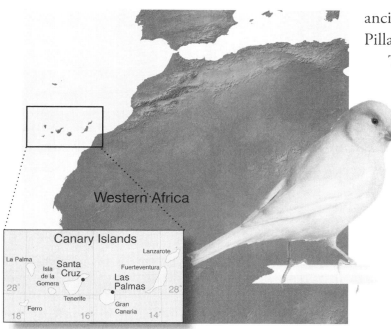
A type of bird, a finch, lived on the Canary Islands. It was a small songbird with greenish yellow feathers.

about 700 miles southwest of Spain and near the western coast of Africa. In sailing days, ships that left the Mediterranean Sea through the Strait of Gibraltar traveled south to catch the prevailing winds to take them west. The islands made a convenient stopping place to take on supplies and check their rigging before making the difficult and dangerous trip across the open sea.

The ancient world was centered on the shores of the Mediterranean Sea. The sea had predictable winds, small tides, and gentle currents. The Mediterranean Sea opened to the Atlantic Ocean through the narrow Strait of Gibraltar. In ancient times, the strait was known as the Pillars of Hercules.

Sailing outside the Pillars of Hercules and into the Atlantic Ocean was filled with dangers. A ship could easily become lost in the vast ocean. Egyptian, Greek, and Roman sailors stayed close to home. But the Phoenicians (fuh-NEESH-uhnz), the great sailors of the ancient world, bravely ventured beyond the Pillars of Hercules.

The Phoenicians were rumored to have discovered islands in the Atlantic. According to the Roman scholar Pliny, a large number of wild dogs roamed the islands. Pliny called the islands Isulae Canariae, or island of dogs. Canariae meant "dog." In English, the name became Canary Islands.

Following the collapse of the Roman Empire in A.D. 476, knowledge of the Canary Islands faded from memory. But in the 1300s, Spain and Portugal began a determined quest for a sea route to the rich spices in China, India, and Japan. The Canary Islands were rediscovered. In the 1400s they became Spanish territory.

The Canary Islands are ancient volcanoes. The largest one rises to more than two miles above the surface of the sea. The islands are noted for their scenery and mild, dry climate. In some places, the coasts are rocky and lined with cliffs, but in other places, they have beaches with golden brown sand. Date palms and cacti grow near sea level, but higher up on the mountain slopes are laurels, holly, pleasant-smelling eucalyptus, and the fragrant scents of flowering plants.

Dogs lived on the islands when the first settlers arrived. The dogs had powerful-looking heads that were almost as wide as they were long. Their necks were thick, and they carried themselves on well-muscled legs. Because they were so strong and fearless, settlers trained them as guard dogs. Although friendly to those whom they knew, the dogs were ferocious toward strangers.

A type of bird, a finch, lived on the Canary

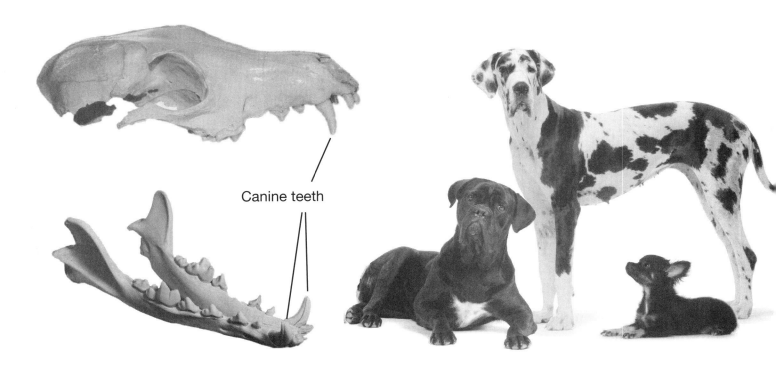

Canine teeth

Islands. It was a small songbird with greenish yellow feathers.

The Spaniards found the volcanic soil to be extremely fertile. Soon the settlers shipped bananas, citrus fruits, peaches, and figs to market. They also grew sugar cane. The small yellow birds traveled to Spain hidden in the cargo. For a time, they were called sugarbirds. However, when the tiny songbird became popular throughout Europe, they became known as canaries.

Canaries were bred to enhance the yellow color of their feathers. They became prized for their showy plumage and their songs. A canary song consisted of bass and flute notes. They could make a sound like bubbling water.

The song of canaries served as an early warning of dangerous gases in mines. Coal miners brought caged canaries with them into the mines. If the birds stopped singing, pockets of dangerous mine gases had collected in the tunnels and threatened to suffocate the birds and the miners.

It is natural to think of the Canary Islands as having been named for the birds. However, the birds were named for the islands, and the islands were named for dogs.

Canine means dog. It is interesting to look for other examples of the word *canine*. The constellation of Canis Major (Big Dog) has Sirius (SIR-ee-uhs) as its brightest star. In fact, Sirius is the brightest star in the nighttime sky. The word *Sirius* is a Greek word meaning "scorching." In ancient times, Sirius shined brightly low on the horizon at dawn during midsummer. The phrase "dog days of summer" refers to the rising of Canis Major during the hot days of summer. Canis Minor (Little Dog) is a nearby constellation. According to ancient mythology, the constellations of Canis Major and Canis Minor represent dogs trotting at the heels of the Greek hunter Orion.

Dogs are carnivores, a word meaning "flesh eating." Dogs and other carnivores have special teeth for holding their prey. These teeth are named after dogs and are called canines. The canine teeth are long and pointed. They help carnivores grasp their prey. In humans, canine teeth on either side of the front teeth on the upper jaw are known as eyeteeth.

Dogs, Canary Islands, canary birds, Canis Major, Canis Minor, and canine teeth all share the word *canine*. Solving the mystery of how names began and what they mean can be fun. But sometimes you have to be a detective and follow the clues to find the answer. Many other names on maps are chosen for birds and other animals. From a map of your home state, see what other interesting animal or plant names you can find.

As another example of how names came into use, consider the guinea pig. The name guinea (GIN-ee) is found worldwide. It is the name of a country in Africa, an island in the Pacific Ocean, a coin from historic England, and several different animals, including the guinea pig, a rodent from South America. Despite being so far-flung, all of the names were derived from the country of Guinea, Africa.

In the 1400s Portuguese sailors explored the west coast of Africa. They named each region by what they found there and shipped back to Europe. Guinea was part of the Gold Coast. Africa also had an Ivory Coast, Grain Coast, and unfortunately, Slave Coast. A part of the Gold Coast became known as Guinea.

The word *Guinea* probably comes from the Berber language. The Berbers lived in Morocco, a country in North Africa across the Strait of Gibraltar from Portugal. The Berbers traveled south by camel caravans. They used the word *guinea* to mean "land of black men," a reference to the native Africans. The Portuguese began using the name, too.

The eastern portion of Guinea next to the sea is gently sloping and covered in part by flat, tropical grassland. The climate is hot and humid. Like many countries near the equator, Guinea has two seasons. One is the rainy season that lasts from June to November. The rest of the year is the dry season.

Sand bars along the riverbanks in Guinea were rich in gold. Fine gold flecks were removed from the sand and shipped to Europe.

The English government minted a coin of gold from Guinea. The gold coin became known as a guinea. The basic monetary unit of the United Kingdom was the pound. In the late 1700s a pound was worth about five United States dollars. A working person earned about 100 pounds a year.

Twenty shillings made a pound. A person

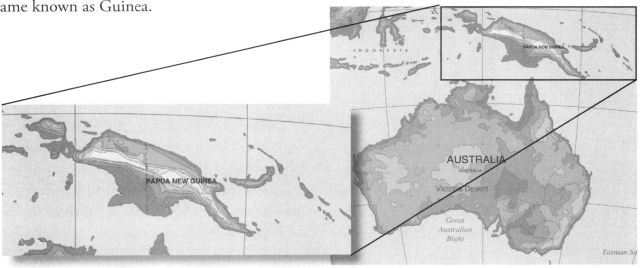

New Guinea

Aristotle

Aristotle (AR-is-totl, 384–322 B.C.) is one of the earliest known persons we can properly describe as a scientist. Aristotle was born in Macedonia in 384 B.C. He grew up in a world where people believed the actions of spiteful gods controlled their lives. If you looked too closely at the world or asked too many questions, these mean-minded gods might punish you.

When Aristotle became 17 years old, his father sent him to attend Plato's Academy in Athens. At the academy, Plato taught that it was all right to ask questions. Aristotle learned to observe, pose questions, and use reason to form conclusions. Aristotle became a teacher at the academy. He stayed in Athens for 20 years.

After Plato died, Aristotle decided to return home. He married, and on his honeymoon trip, he spent carefree hours in the direct exploration of nature. Aristotle especially enjoyed studying animals that lived in the sea. His bride watched as he waded into the warm waters of the Mediterranean. He observed sea life and used a net to catch specimens for further study.

Scholars gave the name "fish" to all sea life. They believed all creatures in the sea were pretty much alike. Aristotle could see differences among the things he caught in his net. He caught an octopus and saw that it did not have scales like a fish. He noticed that dolphins gave birth to live young. This fact, along with other observations, convinced him that dolphins were not fish. They were mammals like horses and cows.

Aristotle wrote about 170 books, an enormous number for one person to write. About 50 of them have come down to us today. The range of subjects and the insights in his writings and lectures is astonishing. He expressed important ideas in astronomy, zoology, geography, physics, and many other subjects, including government.

British shilling

could buy a good meal for one shilling. The value of a guinea coin was set to 21 shillings rather than the 20 shillings that were in a pound. People paid ordinary workers in pounds. But they paid professional people, such as lawyers, in guineas. The lawyers then gave the extra shillings to their assistants.

In addition to gold, Portuguese traded in some of the animals that lived in Guinea. Sailors brought unusual plants and animals to Europe. African plants included guinea corn, the guinea palm, and guinea grass. Animals included the guinea deer and guinea fowl.

Guinea fowl (usually called guinea hens in the United States) came from Guinea. Farmers found the easily excited guinea hens to be especially useful as watchdogs. When a predator came around, the noisy call of the birds warned other barnyard animals and the farmer, too, that something was amiss. They are still used for that purpose today.

A guinea pig has short rounded ears and no tail.

In the 1500s Portuguese ships rounded the southern tip of Africa and explored the Indian Ocean. They discovered an island north of Australia. The island later proved to be the second largest in the world. Greenland is first. The island had a climate similar to that of Guinea, and from the sea it resembled Guinea. For that reason, the Portuguese sailors named the island New Guinea. Guinea and New Guinea are almost halfway around the world from one another.

What about guinea pigs?

Sailors to South America found an animal that lived in big family groups in tall grass. The ten-inch-long animals were stout, short-eared, short-legged, and without any visible tail. Perhaps they reminded the sailors of little piglets. They were shy and easily scared. But when treated gently and spoken to in a soothing voice, the little animals become tame and loving pets.

Sailors brought the South American animals to England and sold them as pets. The country of Guinea was a far-off place with unusual wildlife. The word *guinea* came to mean unusual plants or animals that did not live native to Europe. The little animals from South America became known as guinea pigs.

A guinea pig is not from Guinea, and it is not a pig. It is a rodent. About half of all mammals are rodents. Rodents include mice, rats, squirrels, chipmunks, prairie dogs, groundhogs, beavers, hamsters, and guinea pigs.

The word *rodent* is from a Latin word meaning "to gnaw." The most distinctive feature of rodents is their two front teeth. These are teeth designed for cutting. Guinea pigs have been known to gnaw through aluminum spouts on their water bottles.

A guinea pig's front teeth constantly grow. A guinea pig must gnaw on something to wear them down, even when he is not eating food. Owners who hold and cuddle them will sometimes find that the little animal has chewed a hole in their shirts. The solution is to hold them in an old towel.

At one time, scientists sometimes used guinea pigs as laboratory animals. The term "guinea pig" came to mean an animal, or even a human, that was the subject of research experiments. Today, rats and mice are used instead of guinea pigs.

Today, the main use of guinea pigs is the same as when they were first brought by sailors to England — as friendly pets.

Biologists found that common names for plants and animals changed over time. Canaries were known at first as sugarbirds. In addition, a name could be misleading. The guinea pig was not a pig at all, nor was it from Guinea in Africa or New Guinea in the Indian Ocean. Biologists faced problems in understanding their subject because the same plant or animal could be known under different names, depending on the country or language.

Common names for certain animals can be misleading. For instance, a black panther is the common name for any of several different species of large cats that are black. In Africa a black leopard is called a black panther. In South America a black jaguar is called by that name, too. In the United States a cougar is sometimes called a panther regardless of its color. The cougar is also known as a puma and mountain lion.

Carl Linnaeus (lih-NEE-us), who lived in Sweden in the 1700s, saw that the lack of scientific names hindered the understanding of biology. He proposed a solution that is still used today. Carl Linnaeus was a botanist, a biologist who studied plants. He taught at the Uppsala University in Sweden. His enthusiasm for biology inspired his students to take up the study of nature. When his students graduated and traveled to other countries, they wrote to him about the unusual plants they saw along the way.

He noticed that practically every country had a way of naming plants. Scientists labeled a plant by its common name that changed from country to country. The same plant could have a dozen different names in a dozen different languages.

"If you don't know the name, you waste the knowledge of a thing," he concluded. Carl Linnaeus believed there was order in nature. He valued the Scripture. He believed the Bible when it said that God created all things, including plants and animals.

How could a botanist anywhere in the world

Sunflower

Classification of Mountain Lion

The biological classification goes from the general to the specific. Each plant and animal has a unique name that is shared by no other. The classification for mountain lion, *Puma concolor*, is:

Kingdom: Animal
Phylum: Chordata (animal with a nerve cord)
Class: Mammalia (mother provides milk for the young)
Order: Carnivore (eats flesh)
Family: Feline (cat)
Genus: Puma (an Inca name)
Species: Concolor (one color)

identify a plant regardless of its local name or the language spoken in that country? Rather than a long description, Carl Linnaeus began giving each plant a Latin name based on one of its prominent features.

He introduced his ideas in his book *System of Nature*. The Royal Society of London, England, was the most powerful scientific body in the world. He sent an advance copy of the book to the Royal Society. Botany experts there were reluctant to adopt his methods. They were comfortable with what they already knew. Carl Linnaeus traveled to England to explain what he was trying to do. One by one, he won them over.

In 1751 he published his most important book, *Science of Botany*. In the book, he fully explained how to identify, classify, and name plants. To each plant, he gave a specific, or species, name. The species, Carl Linnaeus believed, reflected God's original plan in the creation of the universe. Next, he collected similar species into a larger group called a genus. The two parts, genus and species, make the name for each plant.

For example, he gave the sunflower the name *Helianthus annuus*, meaning "ring (annuus) of sunlight (Helianthus)." He said, "Who can see the golden blossoms of this plant without thinking of the sun's shape?"

The book made order out of the chaos of botany. Linnaeus laid a firm foundation on which other scientists could build. Scientists could apply his rules for naming plants to animals, too. For instance, all cats belong to the family Feline. Scientists give a separate species name for each kind of cat. The jaguar is Panthera onca; cougar is *Puma concolor*; and leopard is *Panthera pardus*.

Why did Carl Linnaeus succeed? He believed his plan merely showed the order already put in nature by God when He created the universe. Linnaeus always spoke of the species as "the created kinds."

Carl Linnaeus developed a way to name plants and animals so that each living thing would have a unique — one-of-a-kind — name. Another part of biological study is classification.

Classification is the process of putting similar subjects together. A good classification system traces out the order put in creation by God. The Bible describes the world in the beginning as "formless and empty" (Genesis 1:2). The world was in chaos. The word *chaos* means "great disorder or confusion." But God changed chaos into an orderly arrangement. Biologists see the order in nature and make their study of nature easier by grouping plants or animals with the similar features together.

First, biologists separated the living world into two kingdoms — plants and animals.

Animals are capable of moving. They have sense organs to see and hear. They cannot make food energy directly from nonliving minerals. Biologists who study animals are zoologists. *Zoology* is from a Greek word meaning "living being."

Plants have no organs designed to help them move. In general, they are rooted in the soil or float around in water. They do not have well-developed sense organs for seeing or hearing. They have no nervous system. Plants have cell walls of cellulose. They can derive their food energy from nonliving minerals by photosynthesis. Biologists who study plants are called botanists. *Botany* is from a Greek word meaning "plants."

The plant and animal kingdoms contain millions of different forms of life. To be truly useful, a classification system must contain smaller divisions that include similar subjects. Biologists have developed a classification system that begins with the broad category of kingdoms and then into smaller and smaller categories. The next smaller category after kingdom is phylum. The final category is species, or the specific name for a particular animal.

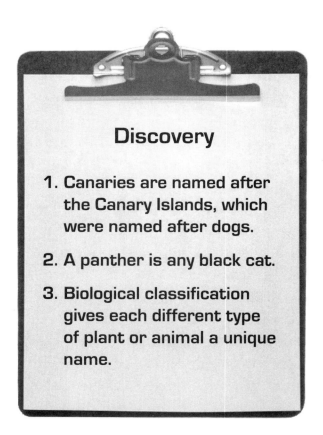

Discovery

1. Canaries are named after the Canary Islands, which were named after dogs.

2. A panther is any black cat.

3. Biological classification gives each different type of plant or animal a unique name.

Questions

1. The Strait of Gibraltar opens from the Mediterranean Sea into the _____ Ocean.
A B C D 2. The first group of sailors rumored to have discovered the Canary Islands were: (A. Greek   B. Phoenician   C. Roman   D. Spanish).
A B C D 3. The Canary Islands were named for: (A. birds   B. cats   C. dogs   D. pigs).
T F 4. When a canary began singing loudly, miners knew the air had filled with dangerous gases.
T F 5. *Canis Major* means hot dog.
6. Another name for eyeteeth is _____ teeth.
A B 7. Aristotle classified dolphins as (A. fish   B. mammals.)
A B C D 8. The English gold coin, the guinea, came from gold found in: (A. Guinea along the west coast of Africa   B. Morocco in Northern Africa   C. New Guinea, an island north of Australia   D. the northern part of South America).
T F 9. Guinea pigs are not pigs but rodents.
T F 10. A guinea pig's front teeth constantly grow.
A B C D 11. The one who wrote *Science of Botany* was: (A. a member of the Royal Society   B. Aristotle   C. Carl Linnaeus   D. Plato).
A B 12. The most general name is Kingdom, but the most specific name is (A. Family   B. Species).
A B 13. The word *chaos* means (A. disorder or confusion   B. orderly arrangement).

Explore More:

The animal kingdom is divided into 33 phyla. Not all of the phyla are well known or contain many species. Explore some of the better-known phyla:

Porifra (sponges)
Coelenterates (jellyfish, hydra, coral, and sea anemones)
Mollusks (clams, oysters, and snails)
Octopuses and squids

  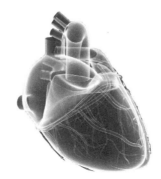 

Chapter 4

Growing a Green World

Some plants, such as sunflowers, can turn to follow the sun. But they do not have a way to move from one place to another on their own. They stay rooted in one spot or float in water. All living things, both plants and animals, require the proper conditions to survive. If the conditions change, animals can move to a better location. Plants, however, cannot migrate.

Plants can be damaged by wind, ice storms, and pest infestation. Great swaths of plants can be destroyed by flood, drought, fire, or changing climate. Although individual plants may die, plants have a design to help preserve the species — new plants can grow from seeds.

The outer part of the seed is usually tough. It holds

Explore

1. How do plants spread their seeds to distant locations?

2. Do plants reproduce solely from seeds?

3. What parts of plants are eaten as food?

moisture inside the seed. It shields the growing part, known as the embryo, from harsh conditions. The seed also contains nourishing food for the growing sprout. A seed that starts to grow is said to germinate (JUR-muh-nate).

Seeds must germinate at the right moment — too soon and they die from freezing weather, too late and they are killed by a hot, rainless summer. Plants usually produce seeds in the fall. But the seeds do not start growing until the time is right. Most seeds delay sprouting until after winter. They remain inactive until they pass through a period of cold weather.

For other seeds, growth is set in motion by rain. Desert plants often use this method. Their seeds contain a chemical that prevents germination. A soaking rain washes the chemical out. For instance, seeds of the desert primrose are dormant until the first rains of spring. Then they sprout and rapidly spread out roots to catch any additional rain. They quickly grow flowers on short stems and produce seeds for the next year.

Producing seeds is not enough. Plants must spread seeds beyond their immediate area. The environment may change and kill the plants. Seeds in the same place as the parent plants would find it difficult to survive, too. In addition, in some locations, the soil has only enough food for a limited number of plants. Seeds need to be spread beyond the growing area of the plant that produced them. Otherwise, they would be competing for food with existing plants of the same species.

Plants have been designed with several different ways to send their seeds beyond the immediate area.

One of the ways is to use animals. Some seeds have an outer coating of burrs and barbs that catch on the hooves or fur of animals. The cocklebur, burdock, and sandburs are examples of seeds with barbs that hang onto animals. The seeds are carried far from their original site. Eventually, the seed falls off or is brushed off.

Some seeds are inside a berry or fruit that acts as a reward to entice animals to eat them. Birds eat the soft part of the berry, but the seed sticks to their beak. Later they clean their beak and knock the seed free. Some seeds have a tough coating that resists digestion and pass through the animal as undigested material. Seeds eaten by animals may fall miles from the original site.

Evergreen Seed Cone

Some plants do not wrap the seed in protective tissue. Examples include pine, spruce, Douglas fir, and cedar trees. These trees are called evergreens because their needlelike leaves stay green all year around. Most evergreens have seed cones. The cone may take two or three years to develop. In some evergreens, the cones open and seeds fall to the ground. In other evergreens, the cones fall to the ground and remain closed. The seeds do not come out until the cone rots away or an animal tears it apart looking for food. Cones of the redwood trees open after surviving forest fires. Seeds are released to replace trees killed by fires.

Human beings help spread the seeds of plants, especially those of fruit and nut trees. For instance, historians believe almond trees were one of the first nut trees to be intentionally grown by humans. Ecclesiastes 12:5 uses almond blossoms as a symbol to show that time is almost up. Almond buds began growing in the fall. When they appear, the harvest is over.

Almond buds grow on the branches in late fall. Cold weather during winter halts the growth. With spring and the arrival of warmer temperature, the tiny buds spring to life. They grow rapidly. The flower petals emerge in a spectacular show of white. After the petals fall, the almonds grow as a nut inside a hard shell. The shell is itself protected by a thick and fuzzy green hull.

Almonds are tasty, brown-skinned nuts. They were first grown for food in China. Today, most of the world's supply comes from the United States. How did the nuts make their way from China to the United States?

Their journey began along the legendary Silk Road from China. More than two thousand years ago, this trade route climbed some of the world's highest mountains, crossing forbidding deserts and endless plains. Traders made the trip because spices from China commanded incredible prices in Europe. Pepper sold for more than its weight in gold.

Almonds came, too. They have high energy content for their weight and size. A handful of the nuts kept a person hiking the dusty trail all

Seeds from the Century Plant

The century plant of the desert southwest spends years preparing for making seeds. The young plant grows close to the ground with spreading leaves with spines along the edge and a spike at the tip. Although it lives in a harsh climate, the plant patiently grows bigger. As the years pass, its leaves grow longer and thicker. The leaves may become six feet long, five or six inches across, and two or three inches thick. The thick leaves are storehouses of food energy.

For 25 years or more it grows, setting aside in its leaves the food energy it will need in a final burst of growth. When the time is right, it begins growing a central stalk, a single spike reaching high into the air. The spike grows at the rate of five to six inches per day. In the space of a few weeks, the stalk grows high over the plant. It is as thick as a man's leg and towers to 30 feet or more.

Limbs holding seedpods the size of dinner plates grow out from the stalk. Birds eat the seeds and carry them away.

New plants must grow from the seeds because building the towering stalk was the plant's dying effort. The century plant uses all of its energy to produce this once-in-a-lifetime seed stalk. It devotes all of the food energy from the roots and leaves to the task. The leaves turn brown and dry. The plant dies.

The word *century* means "100 years." So many years passed before a plant bloomed that pioneers who settled in the Southwest gave the plant the name century plant. However, rather than 100 years, the average plant blooms once every 25 years or so.

Century Plant

Seeds

day long. People who traveled the fabulous Silk Road brought almonds along, not to trade, but to eat on the way. Some spilled along the roadside and began growing. The trees spread throughout countries that bordered the Mediterranean Sea, especially in Spain and Italy.

They made their way to the New World in the 1700s. Almond trees need a hot, dry summer with the rain mostly in the winter. Those conditions are met in California in the San Joaquin Valley. Spanish explorers carried them to the hot, dry climate of California. Today California grows 80 percent of the world's supply of almonds.

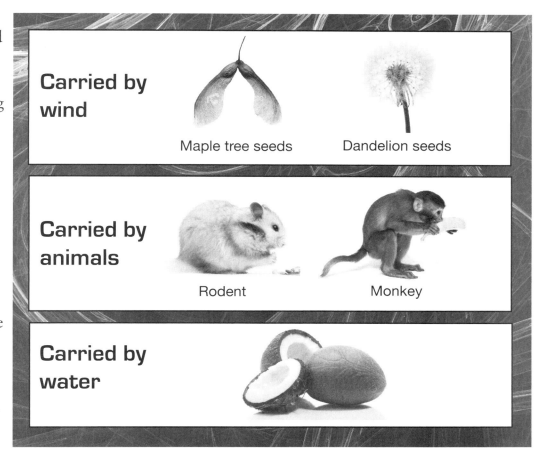

Plants can use wind, water, and animals to distribute seeds. Some plants can shoot their seeds away from the plant.

Some seeds are so light they can hitch a ride on the wind. A dandelion releases seeds hanging from a crown of fluffy hairs. The slightest breeze floats the hairs and propels the seed into the air like a parachute.

Rather than a parachute, the fruit of the maple tree has the shape of helicopter wings. Seeds fall from the tree and whirl in the wind and are carried hundreds of feet from the tree. The seeds are also known as whirlybirds.

Tumbleweeds turn the entire plant into a vehicle to spread seeds. The plant grows as a round bush. After making seeds, it dries up and becomes very lightweight. The stem snaps off and the slightest breeze blows the tumbleweed. As the tumbleweed bounces across the uneven desert landscape, its seeds are knocked loose and fall to the ground.

Some plants use water to carry their seeds. The coconut palm has seeds that are world travelers. The coconut is enclosed in a fiberlike outer husk that helps it float. Coconuts grow along the beach. The big seeds fall into the water, or they roll there. Then ocean currents can carry the coconuts thousands of miles from the original site. Once washed up on shore, the outer husk soaks up rainwater and helps the coconut begin growing.

Cottonwood trees use both wind and water to carry seeds. Their seeds are made of a soft, cottonlike plume. When a grove of cottonwood trees releases their seeds at the same time, the air is filled with cottony seeds like a snowstorm. Cottonwood seeds are so light they can settle on the surface of a flowing stream without sinking. The water transports them even farther than the wind alone.

Other plants are designed to shoot their seeds away from the parent plant. This eliminates the need for them to be carried by wind, water, or animals. Snapdragons have seedpods that dry out. Then the dry pods snap open and hurl their seeds away. The dwarf mistletoe grows high on trees. It shoots its seeds away, and because of its height, they can land 50 feet away. Impatiens (touch-me-nots) also eject their seeds.

Plants can use wind, water, and animals to distribute seeds. Some plants can shoot their seeds away from the plant.

Not all plants reproduce solely by seeds. Many can reproduce by vegetative reproduction. Some plants can grow from part of the parent plant. Animals such as wild pigs dig apart plants, causing both roots and stems to take root and grow into new plants.

Some plants, such as ferns, reproduce by vegetative reproduction. Ferns grow best in warm, damp, and shady places. Rather than true roots, ferns have horizontal stems called rhizomes (RYE-zomes). Most of the stem grows underground. It can send up new shoots along its length. A whole colony of identical ferns grows from the rhizome. The new ferns grow by vegetative reproduction.

Gardeners can duplicate a particular plant by vegetative propagation. They make cuttings from the stem or root of an original and transplant it to a new location. If conditions are right, the root or other small part of the original plant will begin growing. Even plants that produce seeds can be grown in this way. For instance, apple and fig trees can be grown from stems, and grapes can be grown from parts of the vines.

Plants grown by vegetative reproduction are clones. Each one carries the same genetic information as the others, because they all came from a single parent. The daughter plants will be genetically identical to the parent. If a plant has properties that gardeners desire, then vegetative reproduction ensures the daughter plants made from a cutting will have those properties, too.

In the wild, however, plants can be too much alike. Suppose a changing environment kills one plant; then it will kill others, too, because they are identical.

Plants that grow from seeds are not identical. Each trait can come from one of two different genes. Seeds carry traits from the male and female parts of plants to produce offspring. The offspring will not necessarily be exact copies of either one of the parents. Within certain limits, the traits of the offspring will be different from the parents.

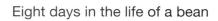
Eight days in the life of a bean

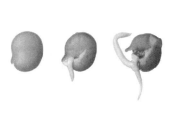

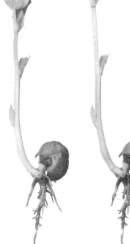

Plant Structure

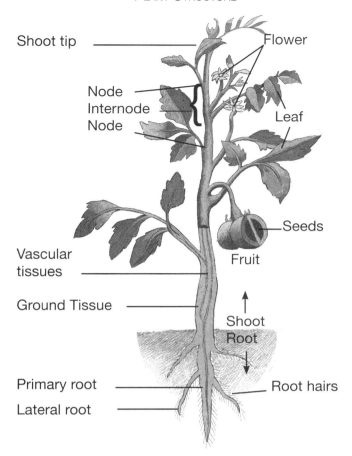

Rather than seeds or vegetative reproduction, some plants can reproduce by spores. A spore is a single living cell protected by a thick cell wall. Plants that do not flower produce spores. These plants include ferns, mosses, and liverworts. They produce spores by the millions. Each spore is light enough to be carried by air currents to other locations. Spores have very little stored food resources compared with seeds. They need very favorable conditions to start growing. The tough little spores can survive for months. But only a few of the millions actually start growing.

Usually, slight variations make no difference in how well plants survive. However, should the environment change, at least some of the next generation may manage to survive the new conditions because they are slightly different from their parents.

In summer, cells grow more slowly and are more closely packed.

In spring, the cells grow faster and are large.

Tree Trunks Are Stems

The trunk of a tree is a stem. The trunk is covered with a protective epidermis (ep-uh-DURM-mus). The word means "outside skin." We call it bark. Bark protects the inner, growing part of the tree. Bark resists the damage of forest fires and thwarts the action of insects. Some bark has a disagreeable taste to keep animals from eating the tree.

Within a trunk are growth rings that are visible when the tree is cut down. Each grown ring represents a year of growth. In spring, the cells grow faster and are large. Their color is light brown. In summer, cells grow more slowly and are more closely packed. Their color is darker. Counting growth rings gives the age of the tree. The distance between the bands reveals whether the year had good growing conditions or poor. Those who study climate use tree rings to learn what weather was like hundreds of years ago.

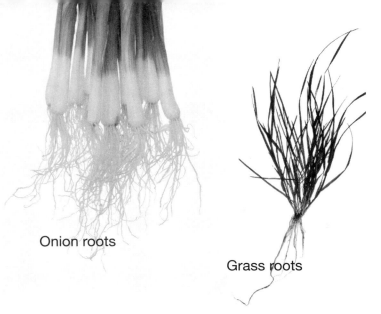

Onion roots

Grass roots

Small tree roots

Plants have many purposes. They provide ground cover and prevent erosion. They serve as homes and places of shelter for animals. They keep oxygen and carbon dioxide in balance in the atmosphere. They also provide food for both man and animals. The Bible explains that the Creator said, "I give you every seed-bearing plant on the face of the whole earth and every tree that has fruit with seed in it. They will be yours for food" (Gen. 1:29).

People are often encouraged to eat vegetables for a healthy lifestyle. The word *vegetable* can have different meanings. A person may use the word *vegetable* as one of three categories: animal, vegetable, or mineral. Some people think of vegetables as those items in the vegetable aisle in the supermarket. Using the vegetable aisle as a guide, green beans, tomatoes, and corn on the cob would be examples of vegetables.

Biologists, however, use the word *vegetable* in a more restrictive sense. A vegetable cannot contain the reproductive part of a plant. Using this definition, many items found at a vegetable stand are not vegetables. Green beans, tomatoes, and corn are not classified as vegetables, because they contain seeds.

Seeds can be grain such as wheat, corn, and oats; seeds can be in fruits, such as apples and oranges; seeds can be the hard pit of a cherry, peach, or plum; seeds can be beans such as green beans, lima beans, and pinto beans; or they can be nuts, such as pecans, walnuts, or hazelnuts. None of these are vegetables in the biological sense.

To biologists, a vegetable cannot be a seed, bean, nut, grain, or fruit. Essentially, then, a vegetable is any food that comes from the root, stem, or leaves of a plant. Examples include carrots (root), asparagus (stem), or lettuce (leaves.)

The root of a plant, found mainly underground, anchors and supports the plant. It draws water and minerals from the soil. Roots differ in how they grow into the soil. Some roots only go down a few inches and then spread out in all directions. Corn has roots that run near the surface of the ground but are wide-spreading. They absorb water that falls as rain. A long period of drought will cause cornstalks to dry and die.

In most areas, water is found several feet below the surface. This source of water, called the water table, is out of the reach of plants with surface roots.

Other plants have roots that seek out water deeper in the ground. Tall trees such as oaks, redwoods, and pines have a main taproot. It goes into the ground as deep as 50 feet. In many places, this reaches the water table. Spreading from the taproot are secondary roots that branch and subdivide until they are as fine as hairs. The roots of a tree are as extensive below ground as the limbs and branches are above ground. Trees can grow long taproots, because their life span is measured in years.

The life span of plants differs depending on how they are designed to produce seeds. Trees and other plants that live for several years before producing fruit and seeds are called perennials.

Some plants, on the other hand, manage to produce seeds in a single year. Corn and wheat grow from seeds, mature, and produce seeds in one season. An annual is a plant that lives its entire life in one growing season. Corn is an example of an annual.

A biennial is a plant that takes two years for its life cycle. During the first season it grows roots, stems, and leaves. During winter, only the roots survive. Next year, stems and leaves regrow, and the plant produces seeds. A carrot is a biennial. During the first year, it grows a thick root full of food. During the second year, the root has the food energy to quickly regrow stems, leaves, and produce seeds. However, people seldom see carrots that have gone to seed, because the root is pulled up and eaten at the end of the first season.

Root vegetables that humans eat include carrots, radishes, sweet potatoes, turnips, and Jerusalem artichokes.

The stem of a plant supports the leaves and provides passageways between roots and leaves for water and nutrients. Stems grow in sections called nodes. Nodes are especially visible in stalks of corn, cane, and bamboo. They are regularly spaced ringlike ridges around the stalk. Leafs and buds grow from the nodes.

Plants whose stems are eaten include rhubarb, asparagus, and white potato. (Celery is part of a leaf stalk rather than the main stem.)

Surprisingly, the potato is a stem vegetable. The eye of a potato is an undeveloped bud. Buds grow from the stem of a plant, so a potato is a type of stem that happens to be found underground.

Potatoes are a major source of food for humans, falling behind wheat, rice, and corn. A potato is a healthy food. It contains only 0.1 percent fat, 3 percent protein, and about 17 percent carbohydrate. The rest is water. Potatoes, however, can be less healthy because of the way they are prepared. They may be deep fried as French fries, baked and then loaded with sour cream and butter, or even made into delicious potato candy.

Carrots (roots)

Asparagus (stems)

Lettuce (leaves)

Coffee

coffee beans

Coffee, a bitter drink that many adults enjoy, is unusual in how it is prepared. A coffee tree produces a berry with a pit. A pit is the hard seed. A cherry is another example of a berry with a hard seed or pit. People eat the berry of a cherry and spit out the pit. But in the case of the coffee bean, the fruit is removed and the pit is kept. Then the pit is dried and roasted. Normally, beans are cooked in hot water, then the water is poured off and the beans are eaten. But in the case of coffee beans, the beans are ground into flakes, hot water is poured over the grounds, the flakes (called coffee grounds) are discarded, and the hot water is drunk. Except for chemicals that soak into the hot water, none of the coffee bean makes it into the coffee.

The leaves of plants also serve as food. Examples of leaves that end up on the dinner plate include cabbage, lettuce, spinach, and turnip greens. Tea is made by soaking tea leaves in water.

Leaves are a food factory. Plants differ from animals in that they can use the energy of sunlight to change minerals into food. Chlorophyll in their cells uses the energy of sunlight to power a series of chemical reactions. The leaf combines water from the roots and carbon dioxide from the air to produce glucose, the sugar that provides energy for living things.

Most, but not all, trees shed their leaves during autumn. About September 22 each year, daylight hours and nighttime hours are of equal length. This marks the start of autumn. As winter approaches, the sun's rays slant at an angle and become weaker. Nights grow longer. Without sufficient light, leaves cannot make enough food. The tree powers down. Sap flows out of the leaves and toward the roots.

After the sap drains from the leaves, the chlorophyll breaks up. The stems holding the leaves become brittle. Wind can break the stems, and the leaves fall to the ground.

The color of the leaves appears to change from green to gold or yellow. Actually, the colors of fall were there in the leaves all along. They were hidden by the much brighter green of chlorophyll. Maple trees put on a great show and turn a brilliant scarlet. Sweet gum, cottonwood, elm, hickory, and walnut reveal leaves of gold, yellow, brown, and orange.

Weather forecasters predict when the colors will be at their peak. Weather conditions do make a difference. Warm, sunny weather with cool nights and the right amount of rain improve the chances of an interesting display of fall foliage.

Discovery

1. Seeds are spread by wind, water, and animals.
2. Plants reproduce from spores, seeds, and vegetative reproduction.
3. The roots, stems, leaves, fruits, and nuts of some plants can be eaten.

Questions

A B   1. Most plants produce seeds in (A. the fall B. early spring.)

A B C D   2. Some plants have seeds with barbs, which are used for what purpose? (A. as a way to be transported elsewhere   B. to protect the growing sprout   C. to provide food for the embryo   D. to delay germination until spring).

           3. What nuts were carried along the Silk Road from China to Spain and then to California? _____.

A B C D   4. What plant has seeds that can be carried thousands of miles by ocean currents? (A. coconut   B. cottonwood   C. dandelion   D. tumbleweed).

T F   5. The thick leaves of the century plant are storehouses of food energy.

           6. The process by which plants can grow from a part of the parent plant such as a cutting from the stem or root is called _____ reproduction.

A B   7. Spores are produced by plants that (A. flower   B. do not flower).

T F   8. A spore is made of a single cell.

           9. Plants that take two years to produce seeds are known as _____.

          10. The white potato is what part of the plant? _____

T F   11. Photosynthesis produces proteins for growing things.

A B   12. The drink made from the hard seed of a tree is (A. tea   B. coffee).

Explore More:

Visit the produce section of a grocery story. Categorize the produce as vegetables or fruits. Do the vegetables come from the roots, stems, or leaves?

Keep a list of the types of plants that you eat each day. Are they vegetables, fruits, seeds, or grain? How are cornflakes and other breakfast cereals made from the raw crops?

Learn the names of the shrubs, flowers, or trees that grow in a nearby park. Identify each one as annual, biennial, or perennial.

What are other uses for plants besides food? Are plants used to make clothing? How is ethanol made from corn?

Chapter 5

Food for Energy and Growth

Plants are one of the main sources of food for human beings. The human body needs food for a variety of reasons. The body needs food for growth, repair, movement, and energy. The source of food for animals and mankind begins with plants. The importance of plants is illustrated by the fact that the Bible describes planting a garden as one of the first things God did after He created Adam and Eve (Gen. 2:8–9).

Cereal grains — wheat, rice, oats, barley, and corn — supply most of the world's food for people and livestock. The advantages of cereal grains as foods are that they grow quickly, provide

Explore

1. What crop did Native Americans use as a staple food?

2. What food can the body quickly use for energy?

3. What food does the body use for long-term storage of energy?

4. What food does the body use for growth and repair?

Food in the New World

When Columbus sailed to the New World (the name given by Europeans to North and South America), most Native Americans lived in small groups. Their daily life matched the climate and land in which they lived. Each morning, men of the tribes along the seashore fished and collected clams and lobsters. Inland tribes hunted deer, turkeys, geese, and ducks.

Native American women harvested berries and nuts. They farmed crops such as beans, squash, and maize, a type of corn. First, they planted maize. After the corn sprouted, they planted beans. The bean vines climbed the cornstalks, which served as beanpoles. Between the rows, they grew squash or pumpkins. The leaves shaded the ground, kept weeds from growing, and held in moisture. Three or four crops grew in one small plot.

Natives in North America made a tea from roots of sassafras. In South America, some of the crops included chili peppers and cacao beans that gave chocolate. The Incas of Peru used the cold nights and hot, dry days in the high mountains to freeze-dry their food. Plants were used for other purposes, too. Indian maidens of the western states perfumed themselves with the smoke from pleasant-smelling prairie grasses.

an abundant supply of food per acre, and can be stored with no processing. Grains keep for years without needing to be frozen, canned, or preserved with chemicals.

More wheat is eaten than any other grain, and about half of all food eaten by humans is wheat, rice, or corn. Corn is not a popular food except in America. People in many parts of the world view corn as only fit for cattle. They ignore corn for human consumption, although corn yields more food per acre than other grains. Corn also has higher energy content per pound compared to other grains. In the United States, however, corn makes many foods, from cornbread to popcorn. Native Americans had been growing corn long before the Spanish and other Europeans arrived. Corn was the main crop for Native Americans, although they farmed other crops as well, such as beans, squash, and pumpkins.

The Spanish called the native crop *maize*, a word that means "corn" in the Spanish language. English-speaking Europeans gave the name Indian corn to the crop from America. The native corn had shorter stalks, smaller ears, and fewer kernels than modern corn. The ears grew about two inches long. Except for size, it was nearly identical to the corn that we grow today.

Indian corn had hard kernels. Usually the natives ground the corn into cornmeal. They shaped thick batter into flat cakes and baked or fried them. Today, cornbread and hushpuppies are two foods made from cornmeal. The kernels of corn could also be boiled to make hominy. Ground hominy is known as grits. In the late 1800s, cornflakes, the breakfast food, were made from grits that had been cooked, flattened thin, and toasted.

Food provides energy to keep the body warm. Humans, as well as other mammals and birds, are warm-blooded. They need heat energy to maintain a set body temperature. The energy comes from chemical reactions with food.

One way to produce heat energy is by oxidation, the chemical combination of oxygen with other elements. As wood burns, oxygen in the air combines with carbon in the wood to give

Popcorn

One of the favorite foods made from corn is popped corn. The word "pop" comes from the sound that the kernels make when heated until they explode.

Kernels of popcorn have soft, moist centers that are covered by a very hard shell. When popcorn is heated to about 400°F, the natural moisture inside the kernels turns to steam. The pressure builds until the kernels explode. The soft material turns itself inside out. One cup of popcorn kernels expands when popped to make about 30 cups of popcorn.

Not all corn is good for popping. Popcorn has kernels with about 14 percent moisture. Some types of corn have kernels so dry they will not pop. Other kernels, such as those grown as cattle feed, have soft spots on top. The steam escapes and releases the pressure before the kernel pops. The type of corn that pops best is called popcorn.

Popcorn is prepared in many ways. Native Americans parched it. They heated sand in a fire and stirred in kernels of popcorn when the sand was fully heated. Another way to prepare popcorn was to put an ear of corn on a stick, dip it in cooking oil, and hold it over a fire. People ate the kernels as they popped.

Popcorn was also used as a decoration. Natives strung popcorn into long garlands to wear as necklaces or headdresses. Spanish explorer Hernando Cortez described popcorn as a grain that, when heated, became a white flower.

In the American Colonies, people popped corn by dropping a few kernels into a kettle with cooking oil, usually lard made from animal fat. After the oil became hot enough to cause the test kernels to pop, more were poured in. The result was a tasty food that had a distinctive aroma.

By the 1900s, popcorn had become one of America's favorite foods. Popcorn was sold in cities and towns across the United States. When mixed with hot molasses, popcorn could be formed into balls. Popcorn balls became the most popular sweet in America, even more popular than candy.

Rather than putting up lemonade stands to earn money, enterprising young people set up popcorn stands. At a nickel a bag or one cent per popcorn ball, they made a good profit.

People who came to the United States from Europe learned of popcorn, and it became popular there, too. For instance, the French sculptor Frédéric Auguste Bartholdi, who designed the Statue of Liberty, visited the United States in 1876. He wrote home about the delicious popcorn, which he had tasted for the first time.

Most of the world's popcorn is grown in the United States. The main popcorn-producing states are Nebraska and Indiana. Because popcorn can be so easily prepared, it is still one of America's favorite foods. The average American eats about 15 gallons of popcorn a year.

Starch is found in potatoes, rice, wheat, corn and foods made from them, such as cereals and pasta.

carbon dioxide. The reaction is so swift it releases heat and light. Combustion is a type of rapid oxidation.

We sit by a fire to stay warm. But when we are outside on a cold day, our body must make its own heat. It does so by oxidation of food. A person who is lost in a snowstorm on a cold night is much more likely to survive by eating food high in energy. Oxidation in the body is slower than the burning of wood, but it produces heat nonetheless.

In addition to heat, chemical reactions provide energy for daily activities. Muscle tissues need energy to do work. Foods the body uses to produce energy are mainly carbohydrates and fats.

The three most important carbohydrates are cellulose, starch, and sugar. Cellulose, starch, and sugar all contain carbon, hydrogen, and oxygen.

Cellulose is the chief building material for plants. Plant cell walls are made of it. The digestive systems of animals cannot utilize cellulose as food. Bacteria can digest cellulose. Cows and termites have these bacteria in their digestive systems. When cows eat grass or termites eat wood, bacteria in their stomachs break down the cellulose so these animals can digest it. Humans do not have the bacteria, so we cannot digest cellulose directly. However, we can eat the meat of cattle that ate the cellulose in grass.

Starch is a type of carbohydrate, but it can be digested. Starch is found in corn, potatoes, wheat, rice, and foods made from them, such as cereals and pasta. Starch has the same atoms — carbon, hydrogen, and oxygen — and in the same numbers as cellulose. However, the individual atoms are linked together differently. Unlike cellulose, humans can digest starch.

Sugar is another type of carbohydrate that can be digested by humans. One type of sugar is sucrose, or table sugar. It comes from sugar cane, sugar beets, or from the sap of sugar maple trees.

Calorie

The value of food energy is measured by calories. A calorie is the amount of heat needed to raise the temperature of one gram of water one degree Celsius. A calorie is a small amount of heat energy, so for measuring food energy, a kilocalorie is used. The prefix *kilo* means 1,000. A kilocalorie, or food calorie, is one thousand times larger than a calorie. The calorie listed on the label of food packages is kilocalorie. A bagel listed as having 150 calories actually has 150 kilocalories or 150,000 calories. In the process of oxidation, it furnishes that amount of heat energy to the body.

An average adult who is inactive can live on about 1,700 calories a day. An active person needs between 1,800 and 2,500 calories. Continuous, hard physical exertion by an adult takes about 3,000 to 5,000 calories a day.

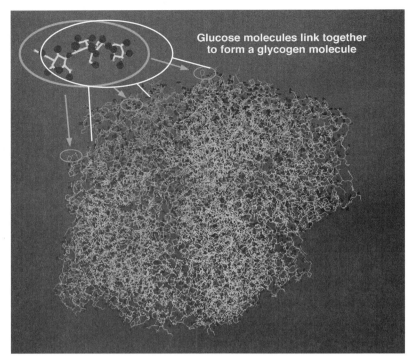

Glucose molecules ($C_6H_{12}O_6$, one magnified at top left) link up to form glycogen. The circles reveal the size of the glucose units within the larger molecule. Glycogen acts as a store of glucose. It is stored in the liver and muscles, a simple sugar that is an important source of energy in the body.

Fructose, another type of sugar, is known as fruit sugar. The sweet taste of ripe apples and peaches is from fructose.

Lactose is a sugar in the milk of mammals. Lactose is called milk sugar. A glass of milk has about a tablespoon of lactose in it. Lactose has only a slightly sweet taste.

Plants such as barley produce a sugar known as maltose. It begins as a starch in the seeds but is broken down into a sugar as the seeds sprout. Maltose is easy to digest. Newborn babies cannot digest cow's milk, but they can digest maltose. It is used to prepare infant formula and make malted drinks. Maltose is also known as malt sugar.

Carbohydrates in the form of starch and sugars cannot be directly used by body cells. Starch and sugar must undergo digestion and be converted into glucose, a simple sugar. The glucose molecule has 6 carbon atoms, 12 hydrogen atoms, and 6 oxygen atoms, $C_6H_{12}O_6$. Glucose is transported by blood throughout the body for immediate use by cells. Glucose is known as blood sugar.

Because glucose is ready for use by cells, hospitals feed it directly into the blood stream of patients who cannot eat or digest food. A tube is inserted into a vein and a plastic bag containing glucose and water is hung overhead. The nurse adjusts a valve to provide a continuous drip of the solution into the bloodstream to keep the right amount of sugar flowing. The drip-drip-drip of glucose ensures that the body is fueled at the proper rate.

In addition to carbohydrates such as starch and sugar, another source of food energy is fats.

Fats contain the same elements as carbohydrates: carbon, hydrogen, and oxygen. However, fat molecules have a higher percentage of hydrogen atoms compared to oxygen atoms than carbohydrate molecules. Oxidation of fats by the body produces twice as much energy as the same amount of sugar. Sources of fat include butter, fats of animals, and vegetable oils such as those from cottonseeds, soybeans, olives, and coconuts.

Foods can provide fat directly, or the body can convert excess sugar into fat. In addition to being oxidized for food energy, fats serve several functions in the body. A layer of fat acts as a cushion to protect some organs from injury. A layer of fat below the skin insulates against cold.

In addition to heat, food provides energy to muscles for daily activities. People who are active in sports must eat foods that provide energy to their muscles. In poor countries, some people cannot take up sports that involve running or bicycling because of the energy required. They cannot afford to buy enough food for the energy they need.

The blood of an adult contains about six grams (about one-fifth of an ounce) of glucose. That is enough to last the body for about 15 minutes. The pancreas, just behind the stomach, regulates the amount of sugar in the blood stream by releasing insulin, a hormone.

A hormone is a chemical messenger. A hormone is produced by one organ of the body and released in the bloodstream to change the activity of other organs. Suppose a person's blood has too much sugar. The pancreas detects this fact and releases insulin.

Insulin in blood causes the liver and muscles to set aside some of the sugar as glycogen (GLY-kuh-juhn). Glycogen is a long, chain-like molecule that the body can quickly and efficiently break apart and change back to sugar when needed.

Processed food and snacks often contain glucose or other sugars that require very little action for the body to use them. After eating a food with a lot of simple sugars, the energy reaches the blood stream in a rush. This gives a burst of energy. Because there is no follow-up of carbohydrates to maintain that energy, the energy level soon drops. Foods that provide quick energy but little else are referred to as having empty calories. Some people call them junk foods.

For the first time in most of human history, people in the United States face an unusual problem. Rather than too little food and suffering from famine, they enjoy a ready supply of a variety of foods. In the United States, the problem for many people is eating so much food that it ends up being stored as fat.

When a person eats too much food, the excess glucose is changed into fat for long-term storage. Fat is the most concentrated form of food energy. When the body needs food energy, glucose is used first. Next is glycogen. Finally, the body begins using stored fat.

Protein-rich food sources

Roasted lamb

Peanuts

Eggs

Fish

Beans

The body is quick to change excess sugar into fat, but reluctant to change fat back into sugar. This is the reason some people find it difficult to lose weight. Rather than trying to lose weight, it is better not to gain excess weight in the first place.

The three main food groups are carbohydrates, fats, and proteins. The primary use of carbohydrates and fats is to give the body energy. The primary purpose of proteins is for growth and repair of the body. The name *protein* comes from a Greek word meaning "of first importance." Proteins are especially needed as children grow into adults. Adults need proteins, too. Hair and fingernails grow and break off, skin cells flake off, and cells in the body die. The daily wear and tear on the body requires a fresh supply of protein.

Proteins are complex compounds made of carbon, oxygen, hydrogen, and nitrogen. Other elements, such as sulfur, are sometimes present. Unlike sugars, proteins are gigantic molecules. Glucose has but 24 atoms, but some proteins contain more than 100,000 atoms.

About half of the solid part of the human body, such as the skeleton, is protein. The body has a huge variety of different protein molecules. Each one has a particular function.

Proteins are found in muscle tissues. They are the strong filaments that contract within the muscle. Proteins also make up enzymes that speed chemical reactions, hormones that act as chemical messengers, and antibodies that fight bacteria, viruses, and fungi that invade the body.

Keratin is a versatile type of protein. It is tough, flexible, and resists wear. Fingernails and hair of humans are made of it. Horns, hoofs, and fur of animals are composed of keratin, as are the beaks and feathers of birds.

Collagen is a protein that makes the connective tissues of the body. Tendons are a band of connective fiber that anchors muscles to the bone. Tendons can flex and absorb shock. Ligaments are larger sheets of connective fiber that attach to bones or hold organs in place. Cartilage, also known as gristle, gives shape to the ears and the front part of the nose. It is more extensive in the skeleton of babies and makes them more flexible than adults. As infants age, cartilage changes to bone. Tendons, ligaments, and cartilage are types of collagen protein.

The body digests protein by breaking it into smaller units known as amino acids. Amino acids are building blocks that the body can put together to make the type of protein that it needs. Scientists have identified 20 amino acids that the body requires. From these 20 amino acids, the body can build more than 30,000 different proteins.

It is possible for the body to use protein for energy when other foods are not available. However, protein is not an efficient source of energy for heat and motion. If the body is so desperate for food that protein must be used, the liver converts protein into an amino acid and a carbohydrate. Only the carbohydrate is changed into energy, and the amino acid is eliminated as waste.

Discovery

1. Corn was a main crop of Native Americans.

2. Sugar, especially glucose, is readily available for energy.

3. Fat stores twice as much energy as other foods.

4. Protein is used for growth and repair of cells.

Questions

A B   1. Most of the world's supply of food comes from (A. protein from nuts   B. cereal grains).

T F   2. Corn was brought to the New World by the Spanish explorers.

A B   3. The body can make heat energy by the (A. oxidation   B. reduction) of food.

A B   4. The energy foods are carbohydrates and (A. fats   B. proteins).

A B C D   5. Starch is a type of (A. carbohydrate   B. fat   C. indigestible cellulose   D. protein).

6. The reason cattle can digest grass is because they have _____ in their digestive system.

A B C D   7. The sugar found in mother's milk is (A. fructose   B. glucose   C. lactose   D. maltose).

8. The simple sugar ready for use by cells is _____.

9. The three elements found in both carbohydrates and fats are _____, _____, and _____ (any order).

T F   10. The body can convert excess sugar into fat.

A B   11. A calorie is a measure of (A. fat   B. heat energy).

A B C   12. The one that the body uses for long-term storage of energy is: (A. sugar   B. fat   C. protein).

A B C D   13. The one used for growth and repair of the body is (A. carbohydrates   B. fats   C. glycogen   D. proteins).

A B   14. The ones that are made of large molecules are (A. sugars   B. proteins).

Explore More:

"Read the label and set a better table" is a ditty that helps people understand what is found in packaged food. In the United States, food packages are labeled with what they contain. Find the food contents label of several different types of food. Notice the serving size. How many servings does the package have? Notice the number of calories per servings. How many grams of fiber and how many grams of fat does it have? Compare those numbers with several other types of foods. Which ones have the fewest calories, the most fiber, and the least fats?

What is a staple food, and what are the most common staple foods? In what countries is rice the main staple food? Are potatoes a food staple? What was the Potato Famine?

Make a list of foods that contain corn.

What are vitamins, and how does the body use them? What minerals and trace elements does the body require?

Chapter 6

Digestion

Carbohydrates and fats are foods that provide energy. Protein is used to build the body and repair tissue. But eating foods that contain carbohydrates, fats, and protein is not enough. Practically none of the foods that we eat each day can be directly used by our cells. Food must be changed into a form that can be carried by the blood and absorbed by cells. Digestion is the process that changes food into simpler substances cells need.

Digestion of solid food takes place in two stages. First the food is crushed and cut into smaller pieces by the teeth. This is mechanical digestion. Then chemical action changes the smaller pieces into basic molecules such as glucose and amino acids. Only then will they dissolve in blood and be absorbed by cells. Babies do not have a full set of teeth and must skip the first step. Their diet is limited to milk and food that is a liquid or nearly so.

Explore

1. Why do human teeth have different shapes?

2. How did biologists learn how the stomach digested food?

3. Why is a sense of taste important?

Teeth

Teeth have a size and shape depending on their purpose. The front teeth of animals that eat meat differ from the teeth of animals that eat grass. Meat eaters have front teeth that are sharp and pointed for cutting flesh. In animals that eat vegetation, such as cattle, the teeth are flat and broad. Their teeth come together to grind and crush grass.

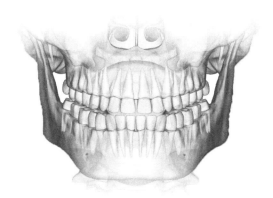

Carnivores — animals that eat meat — have canine teeth, which are designed to grasp and hold prey. Dogs have prominent canine teeth, and in fact, canine teeth are named after dogs.

Humans have teeth that serve a variety of purposes. Human front teeth, called incisors, are sharp and chisel-like. They are designed for cutting and biting. When the jaw closes, the lower incisors fall behind the upper incisors. This gives the teeth a cutting action like scissors.

On either side of the incisors are canines. In humans, the upper ones are sometimes called eyeteeth. Dentists call them cuspids, from a Latin word meaning "pointed." Next are two bicuspid teeth. They have two pointed projections on their biting surface. For that reason, they are called bicuspids. The word comes from *bi* meaning "two" and *cusp* meaning "pointed." They grind and mash food.

Three molars are at the back of each jaw. They have an uneven but generally flat top. The word *molar* comes from a Latin word meaning "millstone." Just as a millstone grinds down grain, so the molars are designed for crushing food.

Human teeth are arranged the same on the left side as the right side, and on the upper jaw as on the lower jaw. We have two incisors on the left upper jaw, two on the right upper jaw, and the same number on the lower jaw, for a total of eight. Each side of each jaw has one eyetooth, for a total of four. Each side of each jaw has two bicuspids, for a total of eight. Each side of each jaw has three molars for a total of 12. This gives a total of 32 teeth.

Young children have but 20 teeth. Baby teeth are missing the bicuspids and the very back molars. Baby teeth are replaced starting at about the age of 7. By age 12, most of the 32 permanent teeth have appeared, except that the very back molars often do not grow above the gums until early adulthood. For that reason, they are called wisdom teeth. They come in when a person has reached maturity and presumably has more wisdom.

Once teeth appear, a greater variety of food can be eaten. Teeth are the main means of mechanical digestion. Chemical changes take place on the surface of food. By chopping the food into smaller pieces, the total surface area increases dramatically. Small bits of food are more easily altered by chemical action.

Saliva begins chemical digestion. Saliva aids in digestion of carbohydrates such as bread, pasta, and other starches. Saliva changes carbohydrates into a type of sugar. This sugar is too complex to be used directly by the cells. Further digestion will change the complex sugar into glucose that cells can use. However,

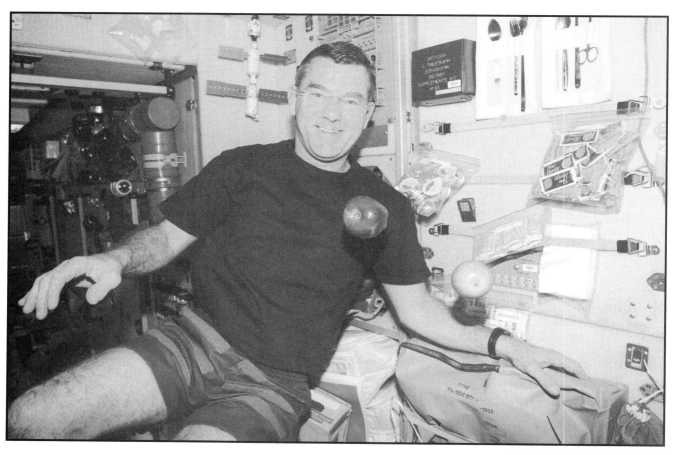

Astronaut James S. Voss, Expedition Two flight engineer, appears to be trying to decide between two colors or two species of apples as he ponders them in the Zvezda Service Module on the International Space Station (ISS).

your mouth can easily taste the change. For instance, put a slice of bread or piece of roll in your mouth and chew it well. The bread will begin to taste sweeter than when you first put it in your mouth.

Chemical reactions of digestion would take place slowly outside the body. Inside the body, however, enzymes make the reactions go very quickly. Enzymes are catalysts that speed chemical reactions. Enzymes are complex proteins formed by living cells. Each enzyme has a particular function. It will act on one type of chemical but not on another one. The enzyme produced by saliva acts on starches.

During eating, the tongue is constantly at work pushing food back between the teeth while avoiding being bitten. The tongue pushes the well-chewed food to the back of the mouth, where it is swallowed. Food enters the esophagus (uh-SOF-uh-guhss), or food tube, that leads to the stomach. Slippery mucus coats the esophagus. The mucus acts as a lubricant to help the food slide along the esophagus to the stomach.

One of the questions that the first astronauts answered was how they would eat in space. Many people assumed that gravity was necessary for food to go down the esophagus. However, muscles that contract in a wavelike motion circle the esophagus. They push the food toward the stomach. An astronaut can eat in weightlessness. In fact, a person can

The process of moving food down the esophagus is called **peristalsis**.

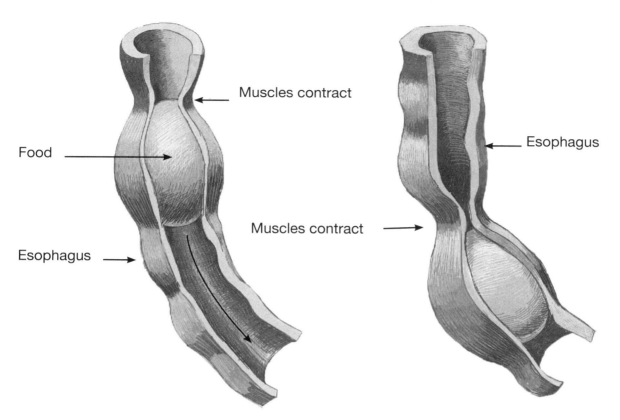

drink, eat, and swallow while standing on his or her head.

Birds do not have these muscles in their throat. Watch birds eat and drink, and you will see them tilt their heads up. They use gravity to help them swallow food. Other birds thrust their beaks forward so the sudden motion sends the food along their throats.

After the lower muscles in the esophagus push food into the stomach, they stay closed. This seals the top of the stomach. A similar muscle at the lower end of the stomach keeps the food in the stomach during digestion. The complete process may take three or four hours, although some foods take longer to digest.

Until the middle of the 1800s, scientists could not agree on how the stomach digested food. Some thought it worked on heat and softened food like a stew pot. Others believed the stomach mechanically mashed the food. An army doctor solved the mystery with the help of Alexis St. Martin, a young man who worked for a French-Canadian fur trapper.

Although stationed on lonely Mackinac Island in Lake Michigan, Dr. William Beaumont kept up with the latest medical research. He was an army surgeon. During the War of 1812, he had one of the highest success rates of saving lives of any doctor.

Gordon Hubbard, owner of the general store, burst into his office. "Come quickly" Hubbard cried. "A man has been shot."

Dr. Beaumont grabbed his medical bag and raced to the general store. He pushed through the men and kneeled by the victim. A fur trapper anxiously wrung his hands. He explained, "My shotgun discharged.

It was an accident. Alexis was standing only a couple of feet away."

The boy had a hole in his chest the size of a man's hand. The sixth rib had been partially blown into the chest cavity. The lower left lung was ruptured. Bone fragments, buckshot, and clothing had entered the stomach.

Dr. Beaumont felt a surge of helplessness. He'd saved the lives of hundreds of soldiers, but this wound would almost certainly prove fatal. Dr. Beaumont cleaned the wound and gave it a dressing. He expected Alexis St. Martin to die at any minute.

The boy managed to continue to breathe despite the massive wound. After 30 minutes, Dr. Beaumont moved the patient to a back room. There he took out fragments of the ribs and pieces of clothing. He treated the wound with a solution of ammonia and vinegar. Doctors had not yet discovered that bacteria cause disease, but the doctor's treatment killed them and prevented infection. Even so, he gave little hope for his patient being alive the next day.

The next day the boy was alive but in grave condition. He had pneumonia, a high fever, and a hacking cough. By the third day, Alexis St. Martin was conscious. Dr. Beaumont continued to treat the wound. After another two weeks, Dr. Beaumont saw that the boy would live, although recovery would be long and difficult.

Alexis was 19 years old and had earned a living as a fur trapper. He could not work with his injury and had no money. Alexis was destitute. Who would pay for the six months of treatment? Dr. Beaumont attended the boy daily at no charge. At first, the French authorities paid for Alexis's room and food. When the French support ran out, Dr. Beaumont and his wife, Deborah, took Alexis into their home.

The hole in the side of the boy's chest refused to heal. The stomach became attached to the skin. A pucker of skin normally kept the inch-wide hole closed. By gently pulling on skin, the doctor could see directly into his patient's stomach. He realized he had an unusual opportunity to study digestion.

Did the stomach merely soften food by heat? Did it mash food? Or was digestion in the stomach a chemical process? Dr. Beaumont had a way to learn the truth.

Alexis agreed to let the doctor examine his stomach and to settle the controversy about digestion.

Dr. Beaumont tied pieces of food to a silk string and put the chunks of food directly into the stomach of his patient. At intervals, he pulled out the food to check the progress of digestion. He learned which foods digested quickly and which ones took more time. He saw that some foods were fully digested in the stomach but others passed on to the intestines. He established that digestion in the stomach was a chemical process.

After two years, the young man was well enough to strike out on his own again. However, he stayed with the Beaumonts as a handyman. He did all kinds of chores, including cutting wood.

Off and on for several years, the army doctor did more than 200 experiments with Alexis. He extracted digestive juices. He sent the gastric juices to doctors and chemists all over the world for study. He found that one of the chemicals was powerful hydrochloric acid.

He noticed how inflamed Alexis' stomach became when Alexis St. Martin drank alcohol. Until then, the army issued a ration of whiskey or rum to soldiers. After these studies, they stopped this practice.

Taste

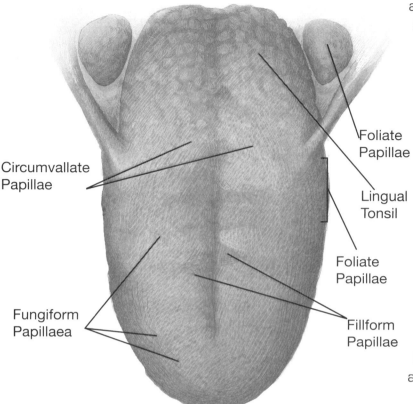

The Tongue
(Papilae are taste buds)

- Circumvallate Papillae
- Foliate Papillae
- Lingual Tonsil
- Foliate Papillae
- Fungiform Papillaea
- Fillform Papillae

The tongue provides a sense of taste through chemical action. Taste is an early warning about the condition of food. Sweet food is usually ripe and safe to eat. Sour or bitter food may not yet be ripe or may be poisonous.

The tongue is covered with small projections commonly called taste buds. The taste buds vary in size. Those on the front of the tongue are nearly microscopic in size and have a smooth, rounded top. Toward the back of the tongue, taste buds are larger in size and rougher.

The four types of taste are sweet, salty, sour, and bitter. Some scientists list a fifth taste, called umami (yu-MAH-mee), for the detection of amino acids such as those in meat broth. A particular type of chemical causes each of these taste sensations. Two foods may be the same size, same temperature, same texture, and look the same. But one will taste different from the other. The difference is due to a chemical reaction between the food and taste buds on the surface of the tongue.

Sweet taste is produced by foods that contain sugar. The human body gets its energy from sugar. When taste buds report that food is sweet, the body responds by desiring the food. Ripe fruits have a pleasant, sweet taste.

Chemicals that become electrically charged when they come apart in water cause salty taste. A salt is a mineral from the nonliving environment. The best-known salt is ordinary table salt, sodium chloride. Table salt is a crystal that dissolves in water. When it dissolves, sodium atoms take on a positive electric charge. Chlorine atoms take on a negative electric charge. Nerves located around the rim of the tongue react to the electric charge and detect it as a salty taste. The body needs salt, but too much salt can cause problems. Because too much salt can pose a danger, the taste buds can detect salt more readily than sugar.

Sour taste is from a substance that contains acid. Vinegar, unripe fruit, lemons, and other citrus fruit contain acids. They taste sour. Sour foods are often not edible. The body is not as capable of digesting sour food as it is sweet food. Taste buds detect the unpleasant and possibly unsafe sour taste far better than the sweet taste.

Bitter taste is from a substance that contains a base, which is chemically opposite to an acid. For humans, bitter foods are often poisonous. The deadly poison strychnine has a bitter taste. The human taste buds are especially sensitive to bitter taste. A little bit of bitterness easily activates them. We are about a thousand times more sensitive to a bitter taste than to a sweet taste. Coffee is one of the few drinks with a bitter taste that some adults enjoy drinking.

Some foods, especially spices, stimulate other nerves in the mouth. The mouth has nerves that report whether a food has a hot or cold temperature. Pepper and mustard activate nerves that detect hot foods. Menthol activates the cold receptors. Our perception of taste is also influenced by the texture of the food, its color, and especially by its smell.

In 1833 Dr. William Beaumont published a book about his experiments. Doctors throughout the world studied it. The book remained the best source of information about digestion for 100 years. Dr. Beaumont retired to St. Louis, where he began a civilian practice. He died in 1853.

Alexis St. Martin lived a full life despite his injury. He returned to Canada and became a farmer. He married and had children. He outlived his doctor and died at age 86.

Dr. Beaumont found that each food is digested at a different rate. When police discover the body of a murder victim, they need to know when the person died. The police call in a medical examiner, who examines the contents of the stomach. He can calculate the time of death based on how far digestion has progressed.

For instance, suppose a meal is made of a glass of milk and a hamburger with meat, pickles, and onions on a bun. Each one digests at a different rate, with the milk going first and pickle taking the longest. The medical examiner can calculate the approximate time of death provided that the time when the victim ate the meal is known.

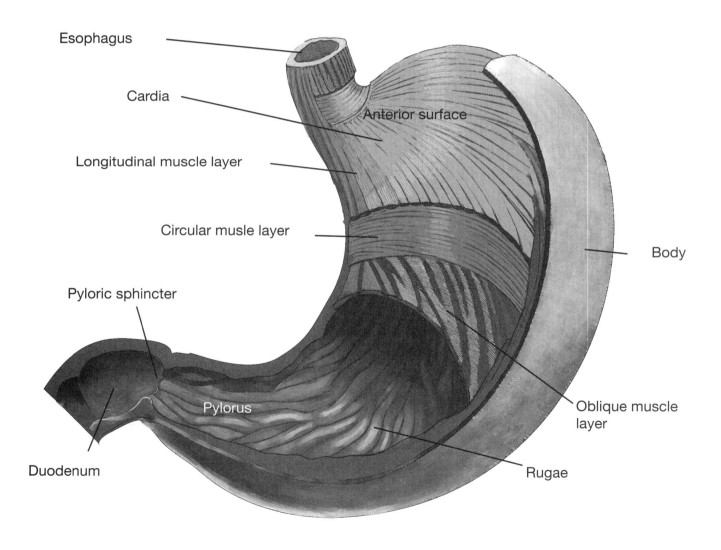

Anatomy of the stomach

Digestion

In addition to juices from the walls of the small intestine, other digestive enzymes enter through tubes from the pancreas and liver. Fluid from the pancreas also neutralizes hydrochloric acid that enters with the food from the stomach.

The liver produces bile, a chemical that aids digestion of fats. Fat tends to collect in globes. In a sense, bile is like a detergent. A person washing dishes uses a detergent to break up grease so water will carry it away. Bile breaks up the larger droplets of fat so they mix better with water. The surface area of a large number of small droplets is greater than the surface area of a single large drop. Because of the greater surface area, enzymes in water more easily combine with small fat droplets.

Most food has been digested after traveling through the first seven feet of the small intestine. The large and complex molecules of carbohydrates, proteins, and fats have been changed into simple compounds that are soluble in water. They will dissolve in blood.

In addition to digestion, the other function of the small intestine is to send food into the blood stream. There is not a direct connection to dump the food into the blood vessels. Instead, it passes through the walls of the intestine.

Absorption is the process by which nutrients pass from the intestine into the blood stream. The wall of the small intestine is lined with small projections called villi. Inside the villi is a rich supply of blood capillaries and lymph vessels. Nutrients in the intestine are separated from the blood vessels by a single layer of cells. Nutrients pass through the single layer of cells by absorption. Glucose and amino acids pass directly into the blood vessels. The fatty material passes into lymph vessels.

The stomach is a J-shaped muscular pouch that can hold about one and a half quarts. The muscles at both ends of the stomach remain closed while the muscular walls of the stomach contract. The purpose of this action is to thoroughly mix the food with digestive juices and enzymes. The juices are released by a multitude of tiny glands located on the inner walls of the stomach.

Pepsin is in digestive juices in the stomach. Pepsin is an enzyme that breaks down proteins into smaller building blocks known as amino acids so they can be used by the body. Foods that are high in protein include meat, fish, eggs, milk, cheese, peas, and beans.

The action of pepsin is improved in a slightly acidic solution. This acid is provided by hydrochloric acid in digestive juices. Hydrochloric acid is made of one atom of hydrogen and one atom of chlorine, HCl. It is a strong acid, but in the stomach, it is diluted. It makes up only about one-half of 1 percent of digestive juices. However, hydrochloric acid is a powerful acid that could eat through the stomach lining. It is prevented from doing so by a mucus lining.

Normally, mucus prevents acid from damaging the tissues of the stomach wall. Occasionally, the mucus lining of the stomach can be damaged. The acid dissolves the mucus lining and exposes the stomach wall. Pepsin attacks protein, including that of the stomach tissues. The pepsin digests part of the stomach itself and produces an open sore known as an ulcer. At one time, doctors thought stress caused ulcers. Research shows a particular type of bacteria causes ulcers.

Once food is reduced to semi-liquid in the stomach, the muscle at the lower end of the stomach opens. The liquid food passes into the intestines. The intestines are divided into two portions, the small and large. The word small refers to the diameter of the intestine, not its length. The small intestine connects to the stomach and is about 1.5 inches in diameter but is coiled around so that it runs for 20 feet. The large intestine is about 2.5 inches in diameter and about 5 feet long.

The small intestine is the main digestive organ. The design of the body is such that it has built-in redundancy. We can breathe through the mouth as well as the nose. The digestive system is redundant, too. Although both the mouth and stomach have a role in digestion, the small intestine can do the complete job for the mouth or stomach. Most digestion takes place in the small intestine. If a person suffers injury or disease to the stomach, it can be removed and the person will survive. The person will have to eat smaller meals more often and watch his or her diet.

Digestion of carbohydrates begins in the mouth, digestion of proteins begins in the stomach, but chemical digestion of fats does not begin until the food reaches the small intestine.

The enzymes in the small intestine not only digest fats, but also give carbohydrates and proteins their final form. The digested form of carbohydrates is glucose. Proteins are changed into amino acids. The final form of fats is fatty acids and glycerol.

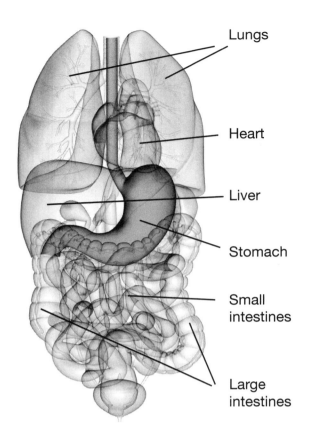

Discovery

1. **Human teeth are designed to cut, hold, and crush food.**

2. **Dr. Beaumont experimented with a patient who had a hole in his stomach.**

3. **Taste reveals whether foods may be good, spoiled, or poisonous.**

Questions

A B   1. Teeth are the primary method of (A. chemical   B. mechanical) digestion.

A B   2. Animals with teeth that are broad and flat are probably (A. meat eaters   B. grazing animals).

A B C D   3. Molar teeth are designed to: (A. cut food   B. grasp food   C. mash food   D. taste food).

A B   4. Saliva changes carbohydrates into a type of (A. sugar B. protein).

T F   5. A person can drink, eat, and swallow while standing on his or her head.

A B C D   6. The stomach digests food by: (A. electrical impulses similar to a microwave   B. chemically changing the food   C. heating it and softening it   D. mechanically mashing food).

T F   7. All foods are digested at the same rate.

A B   8. The acid found in the stomach to help the action of pepsin is (A. hydrochloric acid   B. sulfuric acid).

A B   9. The one that is longer is the (A. large intestine   B. small intestine).

A B C D   10. The chemical digestion of fats begins in the: (A. large intestine   B. mouth   C. small intestine   D. stomach).

A B C D   11. The digested form of protein is: (A. amino acids   B. enzymes   C. fatty acids   D. glucose).

A B C D   12. Taste buds are most sensitive to a: (A. bitter taste   B. mustard taste   C. sour taste   D. sweet taste).

Explore More:

What is lactose intolerance and how is it treated?

What are essential amino acids?

What is a good source of vitamin A, and what condition does the lack of vitamin A cause? Research the same information for vitamins B_1, C, D, and K.

What minerals are essential in the diet of humans? What condition does a lack of iodine in the diet cause?

Why might some birds find it difficult to swallow food in the weightlessness of a space station?

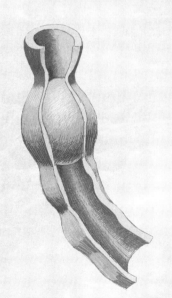
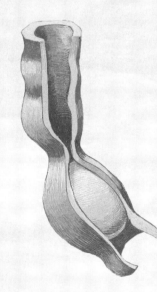

Chapter 7

Plant Inventors

In the 1800s scientists and engineers made dramatic changes to the nonliving environment. They were digging canals, building dams, changing the channels of rivers, and building railroads by blasting tunnels through mountains. They adapted the nonliving environment to better suit the needs of a growing population.

Most biologists did not realize that the living environment could be modified, too. But a few biologists realized that plants could be altered. One of the first to actually do so was Luther Burbank.

As a boy, Luther Burbank worked on his father's farm near Lancaster, Massachusetts. He was the 13th of 15 children. He received a high-school education at Lancaster Academy. His uncle directed a museum in Boston and introduced him to Swiss-American naturalist Louis Agassiz, who encouraged his enthusiasm for nature.

When he grew up, Luther Burbank decided to sell fruits and vegetables that he grew in a 17-acre garden. He faced

Explore

1. How can plants be grown in cold weather?

2. What can be done to reduce an oversupply of certain crops?

3. How do some plants restore nitrogen to the soil?

fierce competition from experienced gardeners who had loyal customers. He planned to succeed by being first to market. But New England had a short growing season because of the cold weather.

Luther built a compost hotbed. The rotting material generated heat, so seeds sprouted and began growing well before spring. Once the weather improved, he set the plants out in the fields. He went to market with sweet corn a week before his competition.

Luther grew other crops, including a red-skinned potato known as Early Rose. He planted potatoes by cutting a potato into sections with an eye in each section. Each eye sprouted to produce a new potato vine.

One day he noticed a potato seed ball. It was a rare discovery. Potatoes seldom produced seeds. Potatoes from seeds were usually inferior to the parent plant. Most farmers grew potatoes by vegetative reproduction. They cut up a potato and planted a piece with an eye in it. The eye sprouted and a new plant grew from it. Luther Burbank carefully opened the seed ball. It contained 23 tiny seeds. The next spring, he planted the seeds. All 23 produced plants.

As gardeners predicted, they were nothing like the original. Some were tiny little potatoes no bigger than marbles. One plant produced misshapen potatoes with sunken eyes. Others went bad almost as soon as he dug them. Others looked great but tasted awful.

One of the potatoes had many desirable properties. It was larger than other potatoes, grew quickly despite harsh conditions, and kept longer without spoiling. The smooth, white-skinned potato tasted good, too, especially when baked.

Three of Luther's brothers who lived in California encouraged him to move there. Their letters convinced him that crops could be grown more easily on the sunny West Coast. To raise money for the trip, he sold the potato to a local grower for $150.

In Santa Rosa, California, he opened a small nursery. He continued to experiment with ways to improve crops. He found three methods that showed promise. First, with artificial selection, he grew a large number of plants but only kept the seeds from those that had the traits he wanted. Second, with cross-pollination, also known as hybridization, he fertilized one plant with the pollen of another to combine two desirable traits. Finally, with grafting, he could graft a good fruit tree sapling onto a sapling that had a strong root system but didn't produce abundant fruit.

Prunes were a highly profitable crop for shipping because they didn't spoil on the trip. One day, Warren Dutton of Petaluma, California, came to see Luther. He was a banker and merchant who had decided to go into the prune business. He said, "I need 20,000 prune trees to plant next spring. The other growers say it is impossible. What do you say?"

Prunes were partially dried fruit from a special type of plum tree. Normally, the trees needed 18 months to grow from seeds. "Let me think about it," Luther said. How could he grow 20,000 trees in 8 months instead of 18 months? The next morning he had worked out a way. He told Mr. Dutton, "You shall have your trees."

He began with almonds rather than prunes. Almond nuts sprouted more quickly than the hard stones of prunes. Almond trees grew more quickly, too. He planted almonds in his greenhouse in a hotbed. He used the trick he had learned in New England of using a hot bed to promote the growth of seeds. After the almond

Love, Faith, and the Apple Tree

John Chapman grew up on a farm in Massachusetts. Each fall, he picked apples from the orchard around the farmhouse. Apples were easy to keep throughout winter. They could be sliced and dried or cooked to make apple butter. Their juice made apple cider and vinegar.

In 1794 pioneer families began moving west. John Chapman decided to advance ahead of them and plant apple seeds. Rather than a bleak wilderness, they would find apple orchards.

John Chapman traveled over the rugged mountains of the Alleghenies into western Pennsylvania. He collected seeds from the cider presses.

As he walked west, his life fell into an enjoyable pattern. He hiked alone. He carried no gun, hunting knife, or traps to take animals for food. He ate cornmeal mush with wild foods such as herbs, fruits, nuts, and berries. In good weather, he enjoyed sleeping in the open. If the weather was miserable, he quickly built a snug lean-to shelter from fallen logs, dead tree branches, and leaves. Along the way, he looked for fertile soil with the right amount of sunlight and moisture.

Johnny Appleseed

One day near sundown, a pioneer family heard a happy song: "The Lord is good to me. And so I thank the Lord for giving me the sun and the rain and apple tree. And some day there'll be apples there, for everyone in the world to share."

The settlers knew the legendary Johnny Appleseed was coming. John Chapman had earned that name after his years in the wilderness. The father wanted to start an orchard. Although he had no money, he knew Johnny would provide the seedlings in exchange for food, clothing, or cornmeal.

In the log cabin, Johnny took none of the food that was offered until he saw that everyone else had been fed. He drank milk with fresh bread. After he finished, Johnny Appleseed said, "I have some news right fresh from heaven." He read the Beatitudes from Matthew (Matt. 5:3–12). He always carried his Bible with him.

With the meager income from the sale of his trees, Johnny bought books on nature, history, and the Bible. When he finished a book, he left it in the care of a pioneer family. The next time he passed through, he retrieved the book and left another in its place. People were eager for learning.

In 1834 he pushed beyond Ohio and traveled into Indiana and Illinois. He kept a chain of orchards in various stages of development spread along three hundred miles. Each year he walked more than a thousand miles.

Fifty years after he started, John Chapman's original goal had become a reality. The fragrance of apple blossoms greeted each new spring throughout western Pennsylvania, Ohio, northwestern Virginia, Indiana, Kentucky, and Illinois. Despite his age, his hair flowed jet black, his eyes sparkled, and he walked with vigorous intensity.

The fathers and mothers who greeted him in Indiana recalled being entertained by his stories when they were children in Ohio.

In March of 1845 John Chapman sang as he walked along the trail near Fort Wayne, Indiana. "The Lord is good to me. So I thank the Lord for giving me the sun and the rain and the apple tree."

The weather turned cold, night fell, and he became chilled. The William Worth family insisted he spend the night with them. By the evening of the next day, he had died of pneumonia. It was the only time he had ever gotten sick. He was buried with a simple headstone that read, "He lived for others."

John Chapman's gentle nature and generous personality had a lasting impact on American life. Pioneer families modeled their independence, humble Christian nature, and willingness to help one another after his example. Much of the pioneer spirit came from John Chapman.

seeds sprouted, he planted them outside. By July, he had 20,000 almond trees that were about waist high.

Almond and prune trees were both members of the same genus. A graft of a bud from a prune tree onto an almond tree would grow. He bought prune buds from neighboring nurserymen who had full-grown trees. Burbank hired dozens of men to graft the prune buds onto the almond trees. He broke the tops of the almond trees to divert the growth into the limbs with the prune buds. This work took almost four months.

By December he was able to deliver 19,500 of the trees to Mr. Dutton. The next spring, he completed the order. Despite the extra effort, he made a profit from the job.

The success encouraged Luther Burbank to consider ways to invent new plants. Developing new fruit trees usually took decades. Some trees did not produce fruit for five years or more. But he now had a shortcut to mature fruit. Rather than waiting for trees to grow to full size, he grafted the buds onto fully grown trees. In a single season, he could taste the fruit to see if it was acceptable. Each twig at the end of a branch could take a different graft. In one case, he had 500 varieties of fruit growing from a single apple tree.

Luther Burbank began thinking of himself as a plant inventor. A few years earlier, Thomas A. Edison had invented a phonograph that could record and play back sound and music. Alexander Graham Bell had invented the telephone. These inventions were changing the world. Burbank believed his discoveries could make a difference, too.

He decided to breed new plants rather than selling existing ones. He ordered plants from all over the world. He believed fruit grown in different soil and light would have different properties.

California produced three types of plums. Two could not stand the rigor of shipping, and one did not taste good. He cross-pollinated them with plums from Japan and elsewhere. His cross-pollination experiments resulted in 10,000 plum seedlings. He reviewed the plants, marked 50 or so that held promise, and burned the rest. He grafted the buds of those he kept onto fully grown trees.

He did not drink alcohol or smoke. He believed either one would affect his senses of taste and smell that were important in testing new plants.

He was always working on dozens of different plants. He developed a type of cactus that had no spines. He thought it might serve as food for cattle. He developed the large and beautiful Shasta daisy, which was one of his favorite flowers.

In 1893 he issued a 52-page catalog that contained more than a hundred of his fruits and flowers, including plums, prunes, apples, blackberries, raspberries, figs, walnuts, and almonds.

Burbank had no formal scientific training. Scientists came to study his methods but left more baffled than educated. Burbank did not keep records that allowed scientists to repeat his experiments. He had a knack for sensing the plants that would do well, but he could not teach this skill to others. However, his work did convince biologists that the living environment could be modified for the service of human beings.

In 1915 Thomas A. Edison and Henry Ford visited him. He met them as an equal because his work had changed the world as much as theirs. His potato had been imported to Ireland. It resisted the potato-blight fungus that caused the disastrous famine of the 1840s. A substantial portion of the world's food supply came from the plants he had developed or improved.

His friends learned that he could not patent his plants. Anyone could grow and sell them. They petitioned Congress to change the law. In 1930 Congress passed the Plant Patent Act. It protected new varieties of plants and the income they produced. With this law, Burbank would have had the exclusive right to grow and sell plants he had developed.

Luther Burbank, however, had died in 1926. In California, trees are planted as part of the Arbor Day celebration on his birthday, March 7.

Another biologist who made a difference and invented new uses for plant products was George Washington Carver.

George Washington Carver was born to a slave during the early days of the Civil War (1861–1865). Shortly after his birth, bandits kidnapped him and his mother. A Union scout rescued George, but his mother, Mary, vanished, never to be seen again. Moses and Susan Carver, who were white but poor, raised the orphaned George as their own. They gave him their last name. He was exceptionally intelligent, but his frail health kept him from doing heavy farm work. Instead, he did laundry, ironing, and gardening.

Young Carver developed a strong desire to learn more about nature. Later in life, he explained, "I love to think of nature as an unlimited broadcasting system through which God speaks to us every hour, if we will only tune in."

When he grew older, it became clear he deserved a better education than one available in the nearby towns. He struck out on his own. The household skills he had learned during boyhood helped pay for his education. When he moved into a new town, he would open a laundry. He could always earn a small but steady income to pay for his education.

He was almost 25 years old before he completed his high school education. Then he attended Iowa State Agricultural College, where he was put in charge of the greenhouse. In 1896 he received an advanced degree at the top of his class in agricultural science. He became a teacher at the college but left when invited to teach at Tuskegee Institute in Alabama.

The American educator Booker T. Washington had begun the school in Alabama. Booker T. Washington believed African Americans could advance themselves through better education and economic development. Carver decided to work with Booker T. Washington at the Tuskegee Institute. He wanted to help others. "It is service to others that measures success," he said. Carver became director of agricultural research.

Poor farmers needed his help the most. Many lived on small 40-acre plots of worn-out land and barely eked out a living. Farmers in the South depended upon crops such as cotton for a livelihood. As the years passed, the soil lost its ability to grow an abundant harvest.

Carver's experiments showed that if farmers planted the same crop year after year, it robbed the

George Washington Carver (1864–1943), reknowned agricultural chemist, working in his laboratory.

soil of its minerals. Carver saw this firsthand. The plot of land he had been given as a research farm had been planted in cotton for years. The soil had become so worn out it would not grow anything, not even weeds. He and the students hauled mud from a nearby swamp to help restore the land.

Could played-out land be restored without hauling in fresh topsoil or spreading fertilizer? The effort of hauling in fresh soil, even if it could be found, was too much for many poor farmers, who had only a mule and a wagon to work the land. They could not afford the expense of fertilizers.

Cocklebur Invention

Each plant has a special way to spread its seeds. In the United States, the cocklebur has seeds in a burr that attaches itself to animal fur or human clothing. In Europe the burdock is another plant that manages to spread its seeds by clinging to fur or clothing.

Whether from a cocklebur or burdock, the burrs are a bother because they cling so well. Removing them from clothing is a chore. In a dog's coat, they sometimes become so embedded the hair has to be trimmed to free them. For years, people picked off the offending burrs and tossed them aside as a nuisance.

In 1948 George de Mestral and his dog took a hike in the Alps. They returned home covered with burrs. The dog waited patiently for George de Mestral to pick out the burrs. As he worked, George wondered what caused the burrs to hold so strongly.

He selected a burr and studied it under a magnifying glass. He was an engineer and quickly saw what made the burr so effective at hanging on. The burr had spines that curved around to form hooks at the end. The hooks stuck in the loops of fabric.

As he experimented with the burr, he realized he could convert the common nuisance into a useful product. His design called for two pieces of material. One would have stiff hooks like the burrs. The other would have soft loops like the fabric of his clothing. Developing a manufacturing process took several years. Making the loops in the soft fabric gave no problem. Making the hooks took more planning. He finally found that heating nylon loops made them tougher. By cutting the nylon partway down the loop, it produced a hook of the right shape.

At first, George de Mestral thought of his invention as a hook and loop fastener. Later, when he was granted a patent, he came up with a trademark name. He put together velour (a soft fabric) with crochet (French for *hook*) to give Velcro.

Velcro is the best known of the hook and loop fasteners. It is used everywhere, from straps that hold on sandals to patches on the Space Station to keep objects from floating about. It can be pulled apart and closed again thousands of times. It can be made to hold so lightly that a small child can undo his coat. It can be made to seal so firmly it holds metal parts together on aircraft. Even dogs benefit from collars that can be adjusted with Velcro.

Thousands of people have picked cockleburs from their clothing. A person with curiosity, willingness to think, and persistence turned a chance observation into a useful product. The Creator has filled nature with wonderful designs that have been used for our benefit. Many more discoveries await people who see solutions where others see only problems.

The four essential elements that plants needed were carbon, oxygen, hydrogen, and nitrogen. Plants extracted carbon from carbon dioxide in the air. Plants got oxygen from a variety of sources, including water and carbon dioxide. Hydrogen was in water. This left nitrogen, which was plentiful in the atmosphere. About 80 percent of the atmosphere was nitrogen. However, plants could not take it directly from the air. Instead, they extracted it from compounds in the soil.

Nitrogen-fixing bacteria converted nitrogen from the atmosphere into compounds for plant growth. Legumes (LEG-yoomz), such as peanuts and soybeans, had colonies of nitrogen-fixing bacteria around the roots.

The important crops of the South in the early 1900s, cotton and corn, did not support the special bacteria. Carver suggested that farmers rotate their crops: grow cotton one year, but the next year plant peanuts, sweet potatoes, or soybeans.

Carver published a yearly newsletter to inform the farmers about his discoveries. During the summer, he drove into the country in a wagon with a portable classroom. He talked to the farmers and offered advice.

Farmers at the mercy of exhausted land put his suggestions in practice. Instead of a single crop, each year they divided their land into sections and planted two or more crops. The next year they alternated where they planted the cotton. Slowly, the soil recovered.

Starting in the 1910s, farmers who heeded his message were spared devastating losses when boll weevils destroyed their cotton. As the pesky insects became more difficult to control, other farmers tried Carver's methods.

The peanut became the South's second largest crop behind cotton. They succeeded almost too well. Prices for peanuts and sweet potatoes fell because they were so plentiful.

Farmers complained, "Peanuts and sweet potatoes are too plentiful. No one wants our produce anymore. Our crops are not worth harvesting."

To overcome the surplus of peanuts and sweet potatoes, Carver devoted himself to finding new uses for the crops. In his newsletter, he listed ways to use peanuts in dyes, soaps, and as milk and cheese substitutes. He gave recipes so others could make the peanut products, such as peanut brittle and peanut candies. From sweet potatoes, he described more than two dozen products, including dyes, candies, pastes, and molasses. These many new uses for the farmers' crops helped them sell briskly.

He gave God the credit for his discoveries. Carver referred to God as "Mr. Creator." He said, "Most of the things I do are just cooking. These are not my products. God put them here, and I found them."

He rose above poverty to become a world-renowned biologist, yet he always remained humble. He said, "It is not the style of clothes one wears, neither the kind of automobile one drives, nor the amount of money one has in the bank, that counts." What counted, he believed, was how well one helped others. Throughout his life, Carver had a profound religious faith, one to which he attributed his successes.

Discovery

1. **Plants can be grown in a greenhouse or hotbed in cold weather.**

2. **New products can be developed from an excess harvest.**

3. **Legumes have nitrogen-fixing bacteria around their roots.**

Questions

A B 1. While in New England, Luther Burbank took sweet corn to market first because he (A. used a compost hotbed   B. imported them from California).

A B 2. Most farmers plant potatoes by (A. planting potato seeds   B. planting a piece of a potato with an eye in it).

T F 3. Of the 23 potato seeds that Luther Burbank planted, only one had many desirable properties.

4. To what state did Luther Burbank move after leaving New England? _____

5. Prunes are partially dried fruit from a special type of _____ tree.

A B 6. To speed up the development of prune trees, Luther Burbank first grew (A. almond trees   B. prune trees) in a greenhouse in a hotbed.

7. John Chapman is better known as Johnny _____.

8. Why did George Washington Carver do laundry, ironing, and gardening rather than farm work? _____

A B 9. George Washington Carver became an instructor at: (A. the Naval Academy   B. the Tuskegee Institute in Alabama).

A B C D 10. The four essential elements that plants need are carbon, oxygen, hydrogen, and: (A. iron   B. nitrogen   C. phosphorus   D. sulfur).

11. In Carver's day, the important crops of the South were corn and _____.

A B C D 12. Carver suggested farmers plant peanuts and sweet potatoes because: (A. nitrogen-fixing bacteria grew along their roots   B. they had fine roots that prevented soil erosion   C. they produced far more income than cotton or corn   D. they released a chemical that killed harmful boll weevils).

T F 13. George de Mestral thought about what caused burrs to hold so strongly while picking them from his cat named Iris.

A B 14. Velcro is also known as a hook and (A. ladder   B. loop) fastener.

Explore More:

Who was Louis Agassiz and what are some of his discoveries?

Luther Burbank did not study the work of Gregor Mendel on dominant and recessive genes. Research Mendel's discoveries and explain how they could have helped Luther Burbank develop plants with desirable properties.

How does one make a compost hotbed? How does a greenhouse provide a warmer temperature for growing plants? What other ways can plants be grown more quickly?

What are some of the better-known varieties of potatoes? Are some better for baking than others? How do plant inventors today use artificial selection, cross-pollination, and grafting to produce better fruit and nut trees?

Can plants be patented like other inventions? Research the Plant Patent Act and other laws that protect the developer of new plants.

What are some of the products that can be made from peanuts and sweet potatoes?

Chapter 8

Insects

A small child who sees an insectlike creature may be fascinated by it or frightened by it. It could be a big, hairy spider, or a tiny red ladybug. The child is likely to call any small, creepy-crawly thing a bug.

All sorts of small things are called bugs — insects, ticks, spiders, and even centipedes and millipedes. However, scientists use a different name for this collection of small life. Scientists call them arthropods (AR-thruh-pod). The word is made of two parts: arthro meaning "joint" and *pod* meaning "foot."

Pod is found in other words. A photographer uses a tripod to hold a camera steady. A tripod has three (tri) feet (pod.) Another form of pod is ped. A pedestrian is a person who travels by foot. Centipedes and millipedes are examples of arthropods with many feet.

Explore

1. How are legs used to classify arthropods?

2. Why are some insects studied more thoroughly than others?

3. How do butterfly and moth antennae differ?

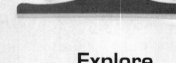

Classification

Kingdom: Animal
Phylum: Arthropod
Class: Insect

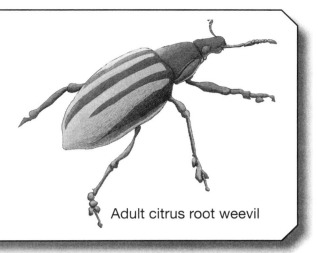
Adult citrus root weevil

The prefex *arthro* means "joint." For instance, a person with the disease of arthritis (ar-THRYE-tiss) suffers from swollen and painful joints (arthri). Always think about a new word to see how it is made. Sometimes you can discover the meaning of an unfamiliar word without looking in a dictionary.

All arthropods — insects, spiders, ticks, lobsters, crabs, centipedes, and millipedes — have feet attached to their bodies by jointed legs. The number of legs identifies them. Insects have six legs, but spiders and ticks have eight legs. Lobsters and crabs have ten legs (counting two claws). Centipedes have one pair of legs for each of their many body segments, while millipedes have two pairs of legs on each body segment.

Insects include bees, ants, wasps, termites, grasshoppers, locusts, crickets, mosquitoes, beetles, butterflies, moths, flies, cicadas, and many others. Insects are found in every type of habitat and climate throughout the world. They outnumber all other animals combined. About a million insects live on a single acre of land. Scientists have estimated that the total mass of all living insects is greater than the total mass of any other animal life.

Entomologists (en-tuh-MOL-uh-gistz) are biologists who study insects. Some insects are beneficial to humans, while others are pests. Some threaten human food sources, but others are of value to humans. Ants, locusts, termites, and other insects, such as mosquitoes and silkworms, directly impact our lives. Entomologists study those insects in detail because they are either a benefit or a nuisance.

Insects that have no obvious role in the daily life of humans are not

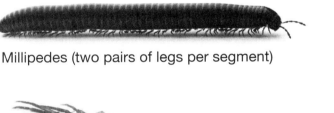
Millipedes (two pairs of legs per segment)

Centipedes (one pair of legs per segment)

Crab (ten legs)

Brown spider and a tick (eight legs)

Locust (six legs)

Jean Henri Fabre Writes to Be Understood

The 1800s were a time of swift scientific change. Before the industrial revolution, what people learned as young workers served them throughout their careers. But rapid scientific advances and technological changes meant workers needed to learn new ways. Government officials in France realized that the education of adult workers was not keeping pace with the rapidly changing workplace. Officials hired Jean Henri Fabre to create an evening course for adults.

However, school officials refused to use his books. They objected that his course material was too entertaining and easy to read. He scoffed at the idea that he could improve his textbooks by making them more difficult to read. He said, "Were I to take their word for it, we are wise only on condition of being obscure."

Fabre resigned rather than change his course material. He began writing nature guides. He wrote with the eyes and mind of a scientist and the heart of a poet. He had an exceptional ability to paint word pictures of his subjects. People could tell one flower from another by his precise descriptions of their aromas and fragrances.

During the period of 1879–1907, he wrote ten science books about insects. He filled the books with direct, patient observations of insect behaviors. The books established him as the foremost authority on insects. Following the scientific books, he wrote popular books such as *The Life of the Caterpillar* and *The Life of the Spider*. He wrote right up to 1915, the year he died at age 92.

His writing impressed scientists such as Louis Pasteur and literary giants such as Victor Hugo, author of *Hunchback of Notre Dame*. Hugo said, "Fabre is the Homer of the insects." (Homer was a talented poet of ancient Greece.)

Fabre wrote books for children, and some of his material made its way into textbooks for French school children. His critics opposed his frequent religious references in the books. Fabre responded, "Without Him (God) I understand nothing; without Him all is darkness. . . . You could take my skin from me more easily than my faith in God."

Fabre became friends with French scientist Louis Pasteur. Like Pasteur, he rejected the idea of spontaneous generation — that life could arise from nonliving matter. He also concluded that the theory of evolution was false. He lived to see most of the main ideas of evolution proposed, but he never accepted the theory of evolution.

---

as well studied. They are often ignored. The total number of insect species may be in the millions, although only about 800,000 have been studied in enough detail to give them names. Much still remains to be learned about the rest of the insects.

In the early days of science, biologists removed insects from their natural environment and studied them in the laboratory. They would capture insects, kill them, and cut them open to see how they were made. They studied not the living animal but the dead insect. They taught students to recognize insects by the three pairs of legs, two antennae, and three body parts: head, thorax, and abdomen.

Jean Henri Fabre, a Frenchman who lived in the late 1800s, disagreed with this approach. Body parts alone did not describe insects. As important to him as their descriptions were their fascinating behaviors while alive in their natural environments.

Rather than looking at dead insects, Fabre preferred going outside and watching the daily lives of living insects. He noticed their wonderful design for finding food. Grasshoppers had mouths designed to eat grass. Beetles had sharp jaws, or mandibles, for catching and devouring insects.

Mosquitoes lived on the blood of animals, so they had a long tube to pierce below the skin. Butterflies had delicate proboscis (proh-BOS-iss) tubes for sucking the nectar from flowers.

Jean Henri Fabre became a schoolteacher. He was fortunate to be assigned to a school on Corsica, the fourth largest island in the Mediterranean Sea. The island rose from sea level to almost 9,000 feet. The change in altitude provided a variety of environments for different plants and animals. Corsica's lush vegetation earned it the name "the scented isle." When wind blew across the island, the fragrance of the flowers could be smelled miles out at sea. Abundant vegetation also assured an abundant insect population.

During his four years on the island, Fabre developed his style for studying nature. He combined patient observations of living insects with detailed but interesting descriptions of their behaviors. He also taught himself drawing so he could correctly depict what he observed.

The usual practice by biologists was to capture an insect, kill, and dissect it. A dead insect held no charm for Fabre. He said, "You rip up the animal, and I study it alive; you turn it into an object of horror and pity, whereas I cause it to be loved. . . . I make my observations under the blue sky to the song of the cicadas."

The classification system developed by biologists puts all arthropods with three body segments into the class Insects. The three divisions of an adult insect's body are the head, thorax, and abdomen.

The head has antennae, compound eyes, and mouth. Biologists are convinced that antennae are sense organs. But do insects use it to detect odors, sounds, or vibrations? Biologists are not certain exactly what the insect uses the antennae to detect.

The thorax, or chest, of an adult insect serves as the attachment for wings and three pairs of legs. Muscles must be anchored on a solid surface so they can pull the legs and wings. Insect muscles are attached to the hard plates of the exoskeleton, a hard outer covering. Wings are made of the same material as the exoskeleton, but it is much thinner and crossed with veins.

The abdomen is the third part of the insect. It has openings, called spiracles, in the exoskeleton for breathing.

Most adult insects have but one goal — to mate and reproduce. Often their lives are short. For instance, mayflies live only 18 hours as adults. Like many other adult insects, they go without food. They don't even have a functioning digestive system. Insects with short life spans as adults have various ways to find a mate. Mayflies tend to all swarm out on the same night and fill the air. Fireflies use flashing lights on their abdomen. Cicadas make a loud buzzing sound.

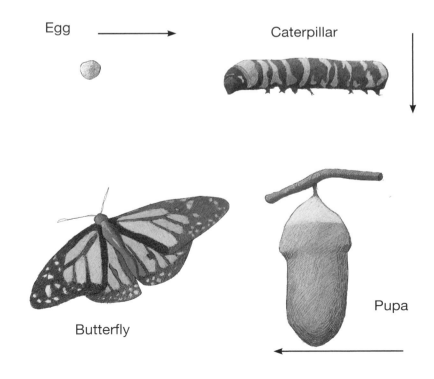

Butterflies have four stages — egg, caterpillar, pupa inside a cocoon, and adult.

A butterfly (left) has a slender body and colorful wings. When it alights, it holds its wings upright over its back. A moth (right) has a thick, hairy body and wings with dull colors. Butterflies tend to fly during the day, but moths fly at night.

Female insects have an egg-laying organ in the abdomen. The organ may also be capable of delivering a sting. The spider wasp uses the sting to subdue a spider. The wasp then carries or pulls the stunned spider to its nest. The wasp deposits an egg on the living but paralyzed spider. After the egg hatches, the larva of the wasp eats the spider. Other insects deposit eggs inside the stems of plants. The egg-laying organ is capable of cutting a slit or drilling into the plant.

The adult stage of an insect may be but a brief part of the insect's entire life span. Insects begin as eggs. After hatching, they may have an appearance that is vastly different from the adult stage.

Insects go through metamorphosis (met-uh-MOR-fuh-siss), a distinct change in appearance. The word *metamorphosis* is from Greek words meaning "to change shape." The change can be profound. Not only can insects change shapes, but they can also change habitats — where they live and what they eat. They may go from living underground as larvae to flying in the air as adults.

In the case of some insects, the transformation is not particularly dramatic. For instance, grasshopper eggs hatch to give a nymph (nimf) that looks like an adult. However, a nymph is wingless and cannot reproduce. As it grows and sheds the exoskeleton, the nymph emerges with wings. At the last stage, it becomes an adult and can reproduce.

Butterflies and moths have a more noticeable change during their metamorphosis. Their life cycles take them through four stages — egg, caterpillar, pupa inside a cocoon, and adult.

Butterflies and moths begin as eggs laid on the underside of leaves. The eggs hatch into caterpillars, the larva stage. The word *caterpillar* is made of two parts: *cat* meaning "cat" and *pillar* from a Latin word meaning "soft, fine hair." The word *caterpillar* means "hairy cat."

At first, the word *caterpillar* ended with the letters "er." But a mistake in a dictionary in 1775 changed the last two letters to "ar." No one caught the error, so the dictionary spelling became the accepted one. Even dictionaries and encyclopedias can contain mistakes.

Butterflies and moths spend their main growing stage as caterpillars. Sometimes caterpillars are called worms, such as the measuring worm, armyworm, and silkworm. The larvae of many insects are efficient eating machines. They can damage crops.

As the larva becomes larger, it must shed its outer covering. It molts. At the last molt, the butterfly caterpillar does not emerge but instead becomes a pupa. The pupa lives inside a cocoon. A cocoon is a case that protects the pupa. Some insects, such as the silkworm moth, spin a silken

Louis Pasteur and Silkworm Disease

In the 1800s, people knew that caterpillars changed forms, but even well-educated people did not know the full story. For instance, in 1865 the French scientist Louis Pasteur was asked to look into the reason silkworms died before spinning their silk cocoons. The manufacture of silk was important to the French economy. Farmers who raised silkworms faced hard times because of the mysterious illness. Louis Pasteur had become France's best-known scientist. The silk growers asked for his help.

Louis believed it was his duty to help the silk growers. He met with Jean Henri Fabre because of his reputation as an observer of silkworms. Louis Pasteur frankly admitted, "I have never so much as touched a silkworm."

Silkworm cocoon

Fabre said, "The silkworms thrive on the mulberry trees and spin their pale gold cocoons in its branches. Silk comes from the cocoons, which the growers soak in a steam bath and unravel to give silk threads. A cocoon is made of a single silk thread 900 meters long." (Nine hundred meters is about 3,000 feet.)

Louis soon learned every phase of the silkworm's life cycle. The silkworm caterpillars became ill before they could spin their cocoons. While examining the tiny dead bodies, Louis found their intestines clogged with masses of microorganisms. Everyone thought the silkworms got sick because of bad air, or something poisoning them, or eating the wrong food. Pasteur proved that small bacteria, or germs, killed silkworms. His discovery led him to the conclusion that germs could cause disease in humans, too. Saving the silkworms of France eventually led to combating human diseases as well.

cocoon. Other insects have a cocoon that is more like paper. The pupa appears to be resting, but enormous changes in its form are taking place.

After the last metamorphosis, the adult butterfly emerges from the cocoon. It has folded wings that it spreads to dry. In an hour or so, the wings are wrinkle free, and the butterfly is ready to fly. The adult has only limited means for getting food. Its mouth does not have biting parts. Instead, it is equipped with a coiled tube for sucking nectar from flowers.

Except for silk from the cocoon, moths and butterflies have few direct economic benefits. Some moths are noted for their destruction as caterpillars because they eat crops or chew holes in clothing. Jesus mentioned moths when He said, ". . . store up for yourselves treasures in heaven, where moth and rust do not destroy" (Matt. 6:20).

The primary indirect benefit of moths and butterflies is that they help pollinate plants. They like to feast on nectar and plant juice. In the process, they carry pollen from one plant to another.

Another benefit, especially of butterflies, is their beauty. Some butterfly pupas are yellow, and many adult butterflies have wings with a buttery yellow color. Perhaps the yellow color gave butterflies

Cicada Broods

Are cicadas seen only once every 17 years? No. They come in broods. Each brood has a 17-year cycle, but different broods come out during different years. A location that sees 17-year cicadas this year may see them again next year, but they are from a different brood. Also, there are 13-year cicadas. They, too, come in broods. The year and location of their surfacing varies. Almost every year a person may be treated to the noisy spectacle. Look for the hollow husks clinging to trees and listen for the unmistakable cicada song.

A brood may cover an entire state, or it may be as small as a few trees in one part of the forest. In 1998 the Midwest was treated to a remarkable sight. The largest brood of the 17-year cicadas and the largest brood of the 13-year cicadas both emerged at the same time. The sound was deafening, and the ground below trees was littered with the discarded husks left by growing adults. The monster cicada swarm will not be seen again in the Midwest for 221 years (17 × 13 = 221).

Cicada

their name. However, no one is sure.

The hobby of butterfly watching is similar to bird watching. People search out new and unusual butterflies, keep a list of those they have seen, and try to photograph them. People go to butterfly houses to see the beautiful insects up close.

Butterflies are noted for their brightly colored wings. Some are iridescent. They shift colors as you move your head. A butterfly wing has overlapping rows of millions of tiny scales. The scales are the size of dust particles. The shimmering color on a butterfly wing is caused by the fine grooves between the microscopic scales. The effect, called *diffraction*, results when light waves interact with one another as they reflect from the fine grooves.

The beautiful wings serve a purpose. Birds feed on butterflies. When a bird gets too close, the butterfly opens its wings. The suddenly display of bright, dazzling colors startles the bird. The butterfly takes the moment of confusion to make its escape.

How do moths and butterflies differ? As adults, butterflies have slender bodies and colorful wings. They fly during daylight hours. When they alight on a flower, butterflies usually hold their wings upright over their backs. Most moths have thick, hairy bodies and dull colors. When they land, they fold their wings next to their bodies. However, some moths look like butterflies.

The main way that moths and butterflies differ is that butterfly antennae end in knobs. Scientists have yet to discover the exact purpose of the knobs.

We normally notice insects flying around in the adult stage. As they fly about, we may not fully realize that this stage of their life is often the shortest. Mayflies are fragile, winged insects that live in water as nymphs for a year or more. Then in May or June, the adults take to the air in great numbers, mate, and usually die in less than a day.

As another example, adult cicadas live about one month as adults but live 17 years below ground as larvae. As adults, cicadas produce a shrill buzzing sound. A large group of them can produce a sound so loud it drowns out a lawnmower. Because of the deafening roar and their vast numbers, cicadas are sometimes incorrectly called locusts. Cicadas are not locusts. Locusts are far more destructive than cicadas.

Locusts are eating machines that can devour entire crops. The Bible says, "What the locust swarm has left the great locusts have eaten; what the great locusts have left the young locusts have eaten; what the young locusts have left other locusts have eaten" (Joel 1:4). Even today, locusts cause much damage in the Middle East. Their location is reported on news broadcasts in much the same way as other countries report the movement of weather fronts.

Cicadas, however, are not locusts. Cicadas are harmless to humans and to crops. After mating, the female finds a tree limb about the thickness of a pencil. She cuts a narrow slit and lays as many as 500 eggs. The ends of the limbs often die and turn brown. The action seldom causes permanent injury to healthy trees.

Adults die within a month. Eggs hatch a few weeks after being laid. The young fall to the ground. The larvae dig below the surface. They are equipped with a mouth like a soda straw that pierces soft tree roots. They drink the sap. As many as 20,000 larvae survive on sap from the same tree. Cicadas have one of the longest life spans of any insect. Most of that time is spent underground as larvae.

After living 17 years below ground, the larvae use their powerful front legs to claw their way to the surface. Larvae crawl up the nearest tree. They fasten their front claws to the bark and quickly change into adults. Their husks split, and adults have about an hour to struggle out of the exoskeletons.

It is a race against time. The flexible adult exoskeleton begins to harden. If the insect fails to get out promptly, it is trapped in the husk. Its wings may harden in their folded position. The defective ones cannot escape predators.

On a warm day in late May, the shrill buzzing of the insects begins. Dry, empty husks of insects cling to the bark of the trees. Large white insects with fierce red eyes and rumpled wings struggle out of the husks. As time passes, the wings straighten and the insects change color to a beautiful brownish yellow.

Cicadas produce a deafening roar. Their distinctive sound comes from tymbals, stiff membranes on either side of the abdomen. Cicadas pop the membrane back and forth so rapidly it produces a steady buzz. A hollow space below each tymbal makes the racket even louder.

Because cicadas live so much of their lives underground and have such a low economic impact, they are not studied as thoroughly as other insects. Much remains to be learned. Why do some follow a 17-year cycle but others a 13-year cycle? Why do the insects make the deafening noise? Some scientists believe it is to confuse their enemies rather than attract a mate. Regardless of the unanswered questions, cicadas are fascinating insects that show the surprising richness of creation.

A ladybug is a beautiful little insect. Its body is round as a ball on top and flat on bottom. The top is bright red with glossy black spots. The head is very tiny, and its legs are short. A ladybug is small; about seven will fit on a dime.

Ladybugs have mouths with jaws that are made for catching small insects. A ladybug is a type of beetle, and like all beetles, it has a strong mouth made for biting. The word *beetle* means "little biting insect." Despite its small size, you can feel the nip of a ladybug bite if you hold it too tightly.

A ladybug is very good at eating other insects, especially aphids. Although aphids are no bigger than pinheads, they are serious plant pests. They can stunt plant growth, cause plants to wilt, and transmit plant diseases. A single female ladybug dines on about 70 aphids a day, while the smaller male ladybug eats about half that many. Ladybugs are a gardener's best friend. They are so popular that five states have made ladybugs the official state insect: Delaware, Massachusetts, New Hampshire, Ohio, and Tennessee.

One legend tells that ladybugs got their name because of how they helped farmers. According to the story, insects were eating crops. Farmers prayed for help. Ladybugs came and ate the pests. The farmers honored the helpful insect by naming it after the mother of Jesus: Mary (Our Lady).

However, this explanation may not be the right one, because during the 1500s, the lady-bug was called a ladybird in England. The word *ladybird* meant "a beautiful young woman." Young men called their girlfriends "ladybirds." The name is still used as a compliment. The wife of Lyndon Johnson (president of the United States from 1963 to 1969) received the nickname Lady Bird when she was a young woman.

During late summer, some ladybugs fly to high mountain peaks and rest on the sunny side of light-colored rocks or buildings. The dramatic color of ladybugs is even more spectacular in the places where they gather. Scientists believe that ladybugs gather in groups to spend the winter. The combined heat of all the bodies helps keep them warm.

One year, astronomers drove to their observatory on the top of Mount Locke in Texas. An astonishing sight greeted them. The white dome of the building that housed their telescope was covered with what appeared to be splashes of red paint. The color came from a layer of ladybugs. Scientists estimated that 35 million ladybugs had flown to the mountaintop.

Merchants visit where ladybugs gather and scoop them up. They sell ladybugs by the gallon to farmers and greenhouse owners. A gallon is about 75,000 ladybugs. The colorful beetles are put in cotton bags with wood shavings. They can be sent to their destinations through the United States mail. The ladybugs are used to control the harmful aphid pests.

Most people can say one version or another of the nursery rhyme that beings "Ladybug, ladybug, fly away home." The original version of the nursery rhyme came from a time when farmers cleared their fields after the harvest by burning the stubble. The original rhyme said:

Ladybug, ladybug, fly away home.
Your house is on fire and your children are gone.
All but one, and her name is Ann,
And she crept under the pudding pan.

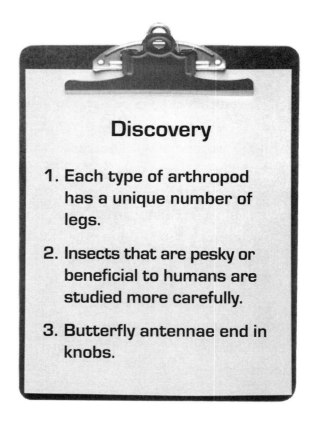

Discovery

1. **Each type of arthropod has a unique number of legs.**

2. **Insects that are pesky or beneficial to humans are studied more carefully.**

3. **Butterfly antennae end in knobs.**

Questions

1. Match the number of legs with the type of arthropod.
   - ___ six
   - ___ eight
   - ___ ten
   - ___ one pair of legs per body segment
   - ___ two pair of legs per body segment

   a. centipedes
   b. crabs and lobsters
   c. insects
   d. millipedes
   e. spiders and ticks

A B 2. The prefix arthro in *arthropod* means (A. foot   B. joint).

A B 3. The one that is a type of insect is (A. cricket   B. shrimp).

T F 4. Biologists often ignore insects that have no obvious role in the daily life of humans.

T F 5. Jean Henri Fabre was the first scientist to bring insects into the laboratory and study them with a microscope.

6. The three divisions of an adult insect's body are _____, _____, and _____.

A B 7. The main goal of adult insects is to (A. eat as much food as possible to survive the winter   B. mate and reproduce).

A B 8. The one with a more noticeable metamorphosis is the (A. butterfly   B. grasshopper).

T F 9. The scientist who discovered the cause of silkworm disease was Jean Henri Fabre.

T F 10. A cicada larva can live underground for as long as 17 years.

A B 11. The one that is a serious plant pest is (A. aphid   B. ladybug).

Explore More:

Although Louis Pasteur, the French scientist, began his career as a chemist, he is today remembered for his biological studies. He found the cause of silkworm diseases and how microorganisms can cause food to spoil. He also discovered how to vaccinate against diseases in animals and humans. Explore some of the diseases he studied. What treatment did he recommend for anthrax, chicken cholera, and rabies?

Insects can carry diseases to humans. What insect carries sleeping sickness? Yellow fever? Malaria?

Bees, termites, and ants are called social insects. What is the meaning of that term? How do social insects differ from other insects? Why do some farmers keep beehives near their crops? Some people say that man's best friend among the insects in the honeybee. What facts support this statement?

Chapter 9

Spiders and Other Arachnids

Both spiders and insects are arthropods. They have jointed legs. But they differ in the number of legs and number of body segments. An insect has six legs and three body segments. A spider has eight legs and two body segments.

Spiders are found throughout the world, on bleak polar ice caps, in wet rain forests, in dry and hot Death Valley, and high on the cold and windy slopes of Mount Everest. They even live under water, although they must breathe air. Water spiders build upside-down silken bowls under water. They collect air bubbles on their furry abdomens and fill their underwater homes with air. Spiders live everywhere humans go, including in the

Explore

1. How do spiders and insects differ?

2. Why was the tarantula portrayed as a threat to humans?

3. Why might a person incorrectly think a pet hermit crab is dead?

84

heart of cities and even aboard airplanes. Biologists have identified about 40,000 species of spiders.

Spiders prey on insects and other small life. They eat flies, moths, crickets, and other spiders. Their fangs inject venom that subdues the prey. The digestive process of spiders starts outside their bodies. They inject digestive juices into the dead animal, liquefy the flesh, and then suck out the liquid contents.

All spiders have poison glands to subdue their prey. But only a few have venom that can harm human beings. Sometimes spiders get a bad reputation, such as what happened to the tarantula. It happened because moviemakers needed a fierce and dangerous-looking spider to project on the movie screen. The early movie cameras were difficult to focus on small spiders.

Filmmakers looked for a large spider to pose as a threat to humans. They found tarantulas to be large, hairy spiders that photographed well. The movie producers portrayed tarantulas as especially dangerous. Tarantulas soon became common villains in motion pictures. They gained a deadly reputation.

Tarantulas *are* big and fierce-looking. The larger ones have a body 4 inches across and a leg span of 12 inches. Tarantulas live in the desert southwest of the United States and in many other tropical and desert climates.

Brown baboon tarantula

They do not spin a web to catch their meals. Instead, they hunt by ambush or by chasing down their prey. Because of their size, they can kill animals as large as mice. But tarantulas bite humans only when threatened, and usually their bite is less painful than that of a bee. Any bite should be treated to avoid infection, but health officials have no record of a tarantula causing human death.

Although tarantulas are not particularly dangerous, two other spiders that live in the United States can harm humans.

A hairy African rain spider

One is the black widow. Female black widow spiders do have bites toxic enough to send some people to the hospital. A black widow's poison is even more potent than that of a rattlesnake, but because of her small size, she cannot inject as much venom as a snake can. Black widow venom attacks the nervous system. Breathing slows, and the person may feel faint. In rare cases, a human can die from the bite of a black widow.

A female black widow spider is only about one-half inch long. She can be identified by a distinctive red hourglass pattern, or figure eight design, on the underside of her black abdomen. The male spider is also called a black widow. He is smaller than the female, seldom seen, and is not a threat to humans.

Cobalt blue tarantula. This tarantula is fast and somewhat skittish. Some specimens have a beautiful and vibrant blue appearance, especially after they molt.

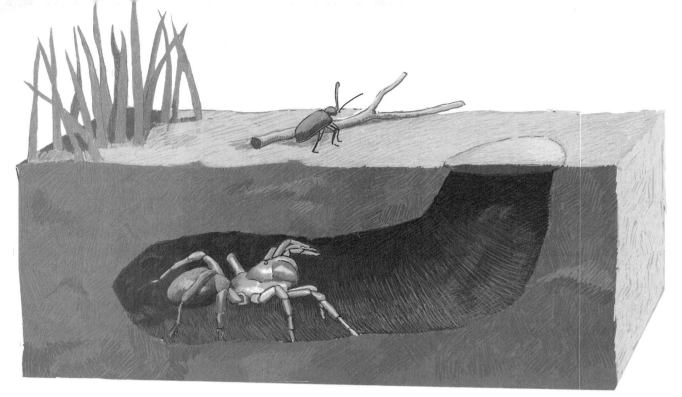

Trapdoor spiders are medium-sized spiders that construct burrows with a cork-like trapdoor made of soil, vegetation, and silk.

The glossy, jet black color is the reason the spider is called a black widow. The widow part of the name comes from the fact that the female sometimes kills her mate. After mating, the female black widow wraps the male in spider silk and kills him. Then she devours him for nourishment.

However, scientists most often see this behavior when they keep black widows in the laboratory where the spiders have limited space to move about. When spider scientists observe black widows in nature, the male is not threatened as often. Perhaps the male is able to escape, or the female turns to other sources of food.

At first, black widows lived in the southern United States. As people traveled, the spiders hitched rides in suitcases. Today, they are sometimes found as far north as Canada. Like most spiders, black widows prey on small life. The spiders make their homes where insects can be found — in woodpiles, tree stumps, and gardens. They like out-of-the-way places in barns, garages, basements, and cluttered places where humans do not bother them.

The brown recluse is the other dangerous spider found in the United States. In fact, most experts believe brown recluse spiders are more dangerous than black widows. Many people cannot identify brown recluse spiders and do not take them seriously when they are recognized. Brown recluses look small and ordinary. They are light brown and have a darker violin-shaped design on their backs. Their bodies are about the size of a dime. Unlike black widows, male brown recluses are as venomous as females. Brown recluse spiders are found mostly in the southern United States.

The word *recluse* means "to withdraw and seek privacy." Most spiders hunt at night, including brown recluses. During daylight, they prefer dark areas and sometimes hide in unlighted corners of homes, in the back of desk drawers, in folds of clothing, or in shoes. Bites often occur because people fail to shake out clothes or shoes before putting them on.

The venom of a brown recluse has a delayed reaction. At first, the bite is nothing but a small red spot. As the venom attacks the walls of blood vessels

Spider Webs in Space

More than 30 years ago, the United States launched Skylab, the first experimental space station, in orbit around the earth. Skylab had 13,000 cubic feet of space, about the same as a three-bedroom family home. With all that room, astronauts could do a variety of experiments.

NASA (National Aeronautics and Space Administration) decided to give students the opportunity to propose experiments for Skylab. Judith Miles of Lexington, Massachusetts, suggested a spider experiment. Spiders would be especially suited for study in space because of their light weight and small size. Spiders quickly adjust to strange surroundings on Earth, but how would they fare in space? She wondered how successful spiders would be at making a web in weightlessness.

Her experiment had a case that held clear plastic panels about two inches apart. The case was about as big as a picture frame. Placed inside the plastic case was a common cross spider that weaves an orb web. The precise, regular spiral of the web would quickly reveal any changes due to a spider having trouble making the web in space.

Arabella, the name astronauts gave to the spider aboard Skylab, proved to be a real space pro. She put smaller webs in each corner of her plastic box. Then she used those webs to anchor a larger one. It looked exactly like the ones she made on Earth.

around the bite, it kills the tissue. A sore develops that grows larger and larger. Flesh can dissolve all the way to the bone. A wound can take months to heal. If infection sets in, the injury can prove fatal.

Despite the danger from black widows and brown recluses, spiders are generally beneficial to humans. They feed on grasshoppers, flies, and mosquitoes. Spiders keep insects under control.

Spiders, like insects, are members of the phylum arthropod. However, spiders have eight legs, two body segments, and eyes that differ from insect eyes. Spiders are arachnids (uh-RACK-nidz). The name comes from a Greek myth. According to this legend, a girl named Arachne became the best weaver of cloth in the world. The Greek gods became jealous of her skill. As punishment, they turned her into a spider to weave forever.

One interesting skill of spiders is their web-building ability and speed at doing so. A spider can

Spider Classification

Kingdom: Animal
Phylum: Arthropod
Class: Arachnid (spider, ticks, mites, and scorpions)

spin a web that spans several feet across a trail. A hiker can walk into the web, destroying it. Yet, an hour or so later on the return trip, the hiker's face once again strikes the sticky web that the spider finished to replace the first one.

Spiders use a sticky web to catch prey. Some spiders, such as orb weavers, make beautiful spiral webs. After finishing its spiral web, an orb weaver hides out on a nearby limb. A signal line runs from the center of the web to the feet of the spider. A struggling insect tangled in the web sends vibrations along the line to the spider. The spider dashes out and stings the insect. It can eat the insect right away or wrap it in silk to save for a later meal.

Spiders use their web material for other purposes, too. Some spiders cast a line for their meals. They let down a line with a sticky ball on the end. An insect is attracted to the ball and becomes stuck to it. The spider reels in the line and eats the insect.

The trapdoor spider has an underground den with a trapdoor cover. The hinge of the trapdoor is made of spider silk. The lid is kept closed until an insect comes nearby. Then suddenly the lid pops open and the spider jumps out to catch its meal.

Many spiders spin a single strand of silk, called a dragline, wherever they go. This dragline is a way to escape when danger threatens. An enemy comes near. The spider drops out of sight down its dragline. When danger passes, the spider crawls back up the dragline.

Young spiders use spider silk to take to the air. During autumn, they climb tall posts and put out long strands of silk. Warm air rising from the ground catches the light silk and carries it aloft. Spiders hold on to the end and are lifted away. They may be carried a half mile up in the sky and come down two hundred miles from their launch point.

As the air cools, the long lines descend and catch in tree limbs and highline wires. At times, the strands from so many spiders cause the air to shimmer as sunlight passes through the glossy strands of nearly invisible spider silk.

The silk of a spider is a wonderful substance. Spider web silk is one of the strongest substances in the world. It is stronger than a strand of steel of the same size. It stretches easily but becomes difficult to break. Spider silk is much tougher than that made by silkworms. They are made in different ways, too. Silkworms extrude the silk through their mouths. Spiders release their silk from spinnerets on the underside of their abdomens. Although silkworm and spider silk are identical in composition, the web-making process of the spider gives a fiber of consistent thickness without weak spots. It does not break as easily.

Today, scientists are using spider silk to make tubes 50,000 times thinner than a human hair. They coat the spider dragline with a liquid that becomes hard after drying. Then they bake the fibers, causing the silk to burn away and the outer covering to shrink even more. The result is a tiny hollow tube that has a variety of uses, including optical paths in extremely small and fast computers.

Biologists classify ticks, mites, and scorpions with spiders because they, too, have eight legs and two body segments. Ticks, mites, and scorpions belong to the class Arachnid and to the phylum Arthropod, same as spiders.

Scorpions live in the southwest desert of the United States. Their tails have a pair of poison glands. They hunt at night, usually by sitting still and waiting for an insect, spider, small lizard, or snake to pass nearby. They sting the prey and then inject digestive juices. Like spiders, they suck the liquefied remains from their victim.

Scorpions also sting when threatened, and their sting can be intensely painful to humans. For that reason, people who live where scorpions are common shake out their shoes and boots before putting them on. They avoid putting their hands into places they cannot see.

Ticks are smaller than scorpions, but certain species pose even greater risk to humans. They are parasites (PAR-uh-sites). The word *parasite* is from

Scorpion

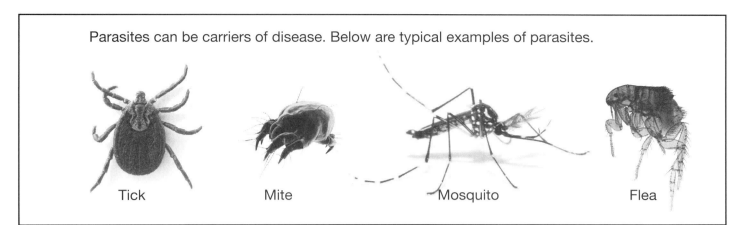
Parasites can be carriers of disease. Below are typical examples of parasites.

Tick   Mite   Mosquito   Flea

a Greek word meaning "to eat at someone else's table." Ticks live on the blood of their host. Ticks discharge a toxin that numbs the nervous system of the victim around the bite. Animals do not feel the tick bite. Some animals, such as deer, carry so many ticks that they die. Ticks are killed by heat. Small brush fires will destroy them. Controlled burns of underbrush by forest service personnel help keep ticks in check.

Tick bites can cause disease in humans. Usually the bite of a tick causes no pain and goes unnoticed because it does not itch. Bacteria can enter through the bite. For a long time, Rocky Mountain spotted fever was the best-known disease caused by bacteria transmitted by ticks. Despite its name, Rocky Mountain spotted fever is found throughout the United States. Symptoms may begin within a day of the bite. A person develops a high fever, headache, and muscle aches. Soon a rash develops that gives the skin a spotted look. Before the development of antibiotics, pioneer families rightly feared the disease, because one person in three who got it died of it. Doctors today have effective treatments, provided the disease is recognized in time.

Lyme disease, not Rocky Mountain spotted fever, is considered the greatest tick-borne threat today. Lyme disease is difficult to recognize, so doctors cannot treat it promptly. Lyme disease was first recognized in Lyme, Connecticut, in 1975. Lyme disease is caused by a bacteria transmitted to people by deer ticks. A skin rash appears, and victims may experience headaches, loss of appetite, and aching muscles. These are symptoms of other diseases as well, such as flu. Patents may fail to tell the doctor about the tick bite because it happened days earlier than the onset of the symptoms. Most victims recover without the cause of the disease ever being known.

In other people, however, the disease may progress to a more dangerous stage. A couple of months after the bite, they complain of pain in their joints. Two years may pass as symptoms become more severe. The nervous system becomes damaged, and a victim may be unable to walk. Treatment is most successful if started as soon as possible.

How can you avoid tick bites? Ticks are too large to work through the weave of tightly woven clothes. A person can wear high-topped socks with pants legs tucked in the socks and long-sleeved shirts. Hikers often check one another for ticks when they stop to rest. A thorough shower within four hours after being outside also helps wash them away before they become firmly attached. Wearing light-colored clothing makes the ticks visible. Tick spray will repel them, too, but it is important to follow directions on the label.

Mites are arachnids similar to scorpions and ticks, but much smaller. They are barely visible to the unaided eye. They can easily work their way through clothing. Some mites live on the flesh of animals, including humans. They attach themselves to the skin and inject digestive juices in the flesh. The chemical dissolves tissue and causes severe itching.

In the southern United States, mites are known as chiggers. Chigger bites can be an unpleasant experience. Scratching the bites can produce sores and open the body to infection.

In addition to spiders, scorpions, ticks, chiggers, and insects, arthropods include a vast number of other animals with jointed appendages and exoskeletons. Of all known animal species, well over half are arthropods. The vast number is due to so many different environments that arthropods inhabit. They live in oceans, freshwater rivers and lakes, on land, and in the air. They range in size from tiny plankton, which blue whales eat by straining them out of seawater, to large lobsters and crabs, which serve as food for humans. Plankton are microscopic in size, while one type of crab is 12 feet across with its legs extended. Both plankton and crabs are arthropods.

One feature common to all arthropods is a hard outer covering, or exoskeleton. An exoskeleton gives support to their bodies, protects their soft, inner body from the bumps and bangs they encounter during daily life, and offers some protection against predators. For sea arthropods, the exoskeleton keeps out water, while land arthropods depend on the exoskeleton to keep them from drying out.

The exoskeleton is made of chitin. It is a strong, rigid compound similar to the material that makes fingernails in humans. It is composed of glucose, a type of sugar, and protein. The chemicals for the exoskeleton come from their environment, so arthropod coverings can have the same color as the environment. It acts as camouflage — a protective coloration that matches the surroundings.

As arthropods grow, they face a problem, because the exoskeleton cannot grow with them. Instead, they must occasionally shed the exoskeleton and grow another one. The process is molting. Molting does have risks. When arthropods molt, they face danger from predators and from the weather. Most land-based arthropods find a comfortable hiding place while the new covering forms.

Building a new exoskeleton takes glucose, which normally supplies energy to the arthropod. Instead, the energy is diverted into making the exoskeleton. Not only is the animal defenseless without its tough covering, but it is also weakened. A hermit crab, which is often kept as a pet, may look so lifeless during a molt that it appears dead.

Arthropods transport oxygen through their bodies by means of blood. Human blood has hemoglobin, a protein that contains iron. Oxygen attaches to the hemoglobin and is carried throughout the body. Arthropods transport oxygen using copper rather than iron. For that reason, their blood is blue rather than red as in humans.

Despite their success, some arthropods have become extinct. Trilobites are one of the best-known examples of extinct arthropods. Their fossils are common, indicating they were once numerous animals on the sea floor. The lenses of their eyes were made of silica, a clear crystal and the chief ingredient of sand. Silica is unchanged by fossilization, so scientists know something about how trilobite eyes worked. No living trilobites have been found to exist today.

Discovery

1. **Spiders have eight legs and two body segments.**

2. **A tarantula was large enough to photograph easily.**

3. **They can appear lifeless during a molt.**

Questions

A B   1. A spider can be described as an arthropod with (A. six legs and three body segments B. eight legs and two body segments).

T F   2. The most deadly spider to humans is the tarantula.

A B C D   3. The spider that has a distinctive red hourglass pattern on the underside of the abdomen is the _____ (A. female black widow   B. male brown recluse   C. orb weaver   D. tarantula).

A B   4. A brown recluse is most likely to (A. hunt during the day   B. hunt at night).

T F   5. Both spiders and insects are arthropods, but only spiders are arachnids.

A B C D   6. Spider silk is stronger than silkworm silk because spider silk (A. contains a small strand of steel   B. has a different composition   C. is consistent in thickness without weak spots   D. is thicker and shorter).

A B C D   7. The one that is NOT a member of Class Arachnid is (A. grasshopper   B. scorpion   C. spider   D. tick).

8. Why might a cowboy who lives in the desert southwest shake out his boots before putting them on? _____

A B   9. The phrase that describes a parasite is (A. to eat at someone else's table B. to live in two different places).

A B   10. The one that is considered the most dangerous tick disease today is (A. Lyme disease   B. Rocky Mountain spotted fever).

11. All arthropods have a hard outer covering known as an _____.

12. Why must arthropods molt? _____

Explore More:

In addition to the number of legs and number of body segments, insects and spiders differ in the design of their eyes. Explore more about the eyes of insects and of spiders and describe these differences.

Lobsters, crabs, shrimps, crayfish, and barnacles have ten legs (counting two claws) and two body segments. Research these arthropods with a goal of describing their benefit to mankind and in what ways they are detrimental. Identify their habitats, their sources of food, and how their designs help them survive in their environment.

Chapter 10

Life in Water

Biologists marvel that the earth is so full of life. Living things manage to grow and thrive in harsh conditions and in unusual places. Biologists estimate earth has about 400,000 plant species. The animal kingdom is far more extensive. About two million species have been studied closely enough to give them names. No one knows how many more animal species have yet to be discovered. Biologists suspect many animal species have not been identified in remote areas of the earth and in the sea.

Animal species do continue to be discovered. In 1997 a small deer never before seen by scientists was found in Myanmar (formally Burma), a country in Southeast Asia. The 25-pound animal stood about 20 inches at the shoulder.

In 2005 biologists in Borneo set up automatic cameras to take pictures at night as animals passed by. Borneo, which has vast tracts of lush rainforest, is a large island in the South Pacific Ocean. The island is home to an abundant

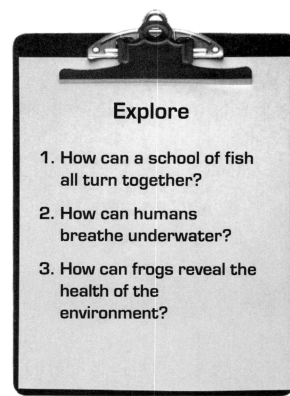

Explore

1. How can a school of fish all turn together?

2. How can humans breathe underwater?

3. How can frogs reveal the health of the environment?

variety of animals. Biologists studied photographs taken by the automatic cameras. The cameras had captured images of an animal they had never seen before. They could not decide on how to classify the animal, because it had features similar to a fox and a cat. But it was clearly an animal previously unknown to scientists.

Fish Classification

Kingdom: Animal
Phylum: Chordate
Subphylum: Vertebrata
Class: Fish

A vast number of smaller living things, such as insects, continue to be revealed in rain forests. An even greater number may be hidden in the ocean depths. Some biologists estimate about five million animal species exist.

In the biological classification system, the largest category is kingdom. The next division is phylum. The plural is phyla. The plant kingdom has but two phyla. Plants with vascular systems for moving sap and nutrients are in one phylum. Plants without vascular systems are in the other.

The animal kingdom is far more extensive than the plant kingdom. It has more phyla, too. Biologists currently put animals in one of 33 phyla. Some of the phyla have obscure animals that are seldom seen, or the phyla have but a few classes of animals. Other phyla have animals that are better known. Sponges are in phylum Porifer. Jellyfish, hydra, coral, and sea anemones are in phylum Coelenterates. Clams, oysters, and snails are in phylum Mollusks. The phylum with the most numerous species is phylum Arthropods, which includes insects and spiders.

The existence of 33 animal phyla can be cumbersome. Some biology students simplify their study by separating animals into two groups: those with backbones and those without backbones. The number of animal species without backbones — invertebrates — is vast. All but one of the 33 animal phyla have nothing but invertebrates. Porifer, Coelenterates, Mollusks, Arthropods, and 28 other phyla are made entirely of animals without backbones.

The only phylum that contains animals with backbones — vertebrates — is phylum Chordata. All chordate animals have a bundle of nerves that runs along their back. This nerve cord is supported by a notochord, a dense protective rod of collagen fiber or cartilage. In some animals, strong vertebrae replace the notochord with a backbone. These animals are called vertebrates — fish, amphibians, reptiles, birds, and mammals.

The biological division below phylum is class. The phylum Chordata has seven classes. Two of the classes — sea squirts and lancets — have a notochord but not a backbone, so they are not vertebrates. Because vertebrates do not make the entire phylum of chordate, vertebrates are considered a subphylum. The subphylum vertebrate is a very small part of the total animal kingdom. Of the two million known animal species, only about 55,000 are vertebrates.

Despite the small number, however, vertebrates include animals that are well known to humans and especially important to us. Fish, amphibians, reptiles, birds, and mammals have an economic impact on our daily lives. Throughout history,

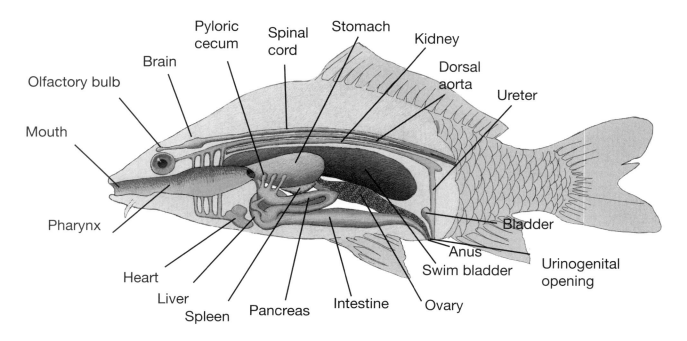

FISH ANATOMY

vertebrates have been used in many ways. The muscles of horses, oxen, elephants, and camels lightened the work that humans had to do. Goats and cattle provided milk. Sheep provided wool for clothing. Falcons were trained to hunt for food. Dogs were trained for hunting and became trusted companions to humans.

Fish are vertebrates that live in water. They are well designed for moving easily through water. They have fins for swimming and scales for protection. They have a streamlined shape to reduce the resistance of water.

Like all vertebrates, fish have a heart that pumps blood and carries oxygen throughout the body. Like land animals, fish require oxygen to survive. But they do not have to breathe air. Instead, gills extract oxygen dissolved in water. Gills of fish play the same role as lungs of land-based vertebrates.

Gills are made of a vast number of very thin membranes. They are folded so that a tremendous amount of surface area fits in a small space. The membranes are filled with a fine network of blood vessels. Oxygen flows from water to blood by passing through the membrane. As oxygen enters blood from the gills, carbon dioxide passes the other way into the water.

Divers have admired the ability of fish to extract oxygen from water. Inventors have tried to duplicate this feat with a special backpack made of thin membranes that lets oxygen pass but keeps water out. However, such a gill backpack to allow humans to breathe underwater has not yet been successful.

The backbone of a fish, like that of other vertebrates, is made of a series of vertebra bones. The vertebra encloses the spinal column, which is part of the nervous system. The nervous system includes the brain, sense organs such as the eyes and ears, and nerves that go from sense organs to the brain.

Fish have the same senses as other vertebrates, including sight, hearing, smell, taste, and touch. Each is important to the fish. For instance, salmon return to the small stream where they were born

Catfish

to spawn. As they swim upstream from the sea, they use their sense of smell to choose which tributary takes them home.

Catfish use their whiskers, which provide a sense of touch and taste, to feed along the bottom of water that is too murky for them to see clearly. Fish can also hear, although they have no external ears. Instead, the sense of sound is located within their skull. The sound is conducted directly to the ear from the water through the skull.

Fish also have a special sense organ, the lateral line, which detects changes in water pressure. Water is a fluid that easily transmits pressure. The lateral line is faintly visible on many fish. It extends lengthwise along the middle of each side. As a fish swims, the action of its tail sends out pressure waves that reflect from obstacles and other fish. The lateral line detects the waves. The fish can avoid an object without brushing against it. Fish that travel in schools turn together because of the pressure wave produced by their neighbors.

Fish are cold-blooded. They have no way to regulate their internal body temperature. Instead, their bodies take on the same temperature as the water through which they swim. Some species of fish survive in water that is only a degree or two above freezing.

Learning about fish in the deep sea was especially difficult for early biologists. Water has weight. It presses against anything submerged in it. At a depth of 33 feet, water pressure is twice that caused by the atmosphere. Fish that live deep in the ocean are designed to withstand the tremendous pressure. But they cannot tolerate the reduced pressure at the surface. When deep-sea fish are brought to the surface in nets, they cannot survive long, even if they are put in water tanks. The water does not have the pressure to keep them alive. In addition, watching a fish in a tank reveals little about its life cycle. Studying the daily life of fish is best done in its normal habitat down in the water.

In 1870 Jules Verne, a French writer, wrote the book *Twenty Thousand Leagues Under the Sea*. In the book, the fictional Captain Nemo and his crew traveled underwater in his submarine the Nautilus. A league is about three miles, so 20,000 leagues refer to the submarine's travels around the oceans and not the depth to which it plunged. In the book, Verne described some of the wonders of the deep. He also described diving equipment that made it possible for humans to walk among the fish. People who read Verne's book became aware of the sea and what it contained. But studying sea life scientifically remained difficult until the invention of the scuba gear by another Frenchman, Jacques Cousteau, in 1943.

Jacques Yves Cousteau was a scientist who opened the undersea world to exploration for both scientists and ordinary people. Born in Saint-André, France, Cousteau received no training as a scientist. After he finished high school, he dreamed not of the sea but of the air. He wanted to pilot an airplane.

Clownfish

Shortly before he started pilot training, he was in a terrible automobile accident. The bones of one arm were broken in five places. The other arm was badly infected. The broken arm healed, but the infected one became nearly immobile.

At Toulon, France, a port city on the Mediterranean Sea, he began taking daily swims to strengthen his arms. By holding his breath, he could dive to 60 feet and stay under for two minutes. Then he had to surface for air. He wanted to dive deeper and stay down longer.

At that time, a diver wore a heavy suit of rubberized canvas and a copper helmet with glass windows. The suit was tethered to a surface ship with a rope and air hose. The surface ship pumped fresh air down the hose to the diver. A diver imprisoned in the suits could barely walk. Swimming freely was out of the question.

Jacques Cousteau experimented with tanks of compressed air. The problem was regulating the flow of the air as it escaped from the tanks. Water pressure increased with depth. The air in the tanks had to be released at a matching pressure. Too little air pressure and the diver's lungs would not be able to expand. He would not be able to breathe. Too much air pressure and the lungs would blow up like balloons.

Cousteau journeyed to Paris and presented the problem to Émile Gagnan, a brilliant engineer. In just three weeks, Gagnan built a valve that regulated flow of air as a diver changed depth.

Cousteau strapped tanks of compressed air on his back. Air from the tanks flowed through hoses to a special mouthpiece. A valve in the mouthpiece adjusted air pressure to equal the pressure of the water. The valve did double duty. It let air into his mouth when he inhaled. It let air escape into the water when he exhaled.

With it, Cousteau could breathe underwater. It was self-contained and did not require a lifeline connected to the surface. He could freely glide through the water. Cousteau made his first successful dives with the Aqualung ("water lung") in 1943. Later, his invention became known as SCUBA gear, meaning "self-contained underwater breathing apparatus."

Cousteau developed ways to take pictures underwater. Designing an underwater camera was difficult. Cameras were not automatic then and used rolls of film. A roll of film had to be cranked forward by hand and the lens focused to the right distance. This was difficult on land, and nearly impossible below water. Undersea cameras required special watertight housings. In addition, water changed where the lens focused. To photograph a fish six feet away, the lens had to be set at four feet. Water also filtered color from the light and gave undersea life a blue cast. Cousteau placed colored filters over the camera lens to restore the missing color.

In 1953 Cousteau published *The Silent World*, a book about life beneath the sea. It was filled with his color photographs. Some oceanographers claimed that his photographs were too beautiful to have been done in the sea. They claimed he must have taken the pictures

Salamander Red-eyed Tree Frog California Newt

in a studio with fish in an aquarium. But he had photographed everything on location underwater.

Part of his worldwide exploration of the ocean was funded by a television series he produced: *The Undersea World of Jacques Cousteau*. Viewers of his television series were treated to underwater views from the Amazon to the Antarctic. They saw schools of fish, sharks, whales, dolphins, sea turtles, and giant octopuses.

By the time Jacques Cousteau died in 1997, he had opened the wonders of the undersea world to exploration for both scientists and ordinary people. More than six million people had taken up diving as a hobby. Millions more had a greater appreciation of the wealth of life in the sea because of Jacques Cousteau's efforts.

Amphibians are another class of vertebrates. Amphibians include frogs, toads, and salamanders. The word *amphibian* is from two Greek words meaning "life on both sides." They are born, develop, and spend their youth in water. As adults, they live on land.

The organs of respiration for amphibians include gills, lungs, and skin. Frogs as tadpoles have gills. Frogs as adults have lungs. They are also capable of absorbing oxygen through their thin skin. A frog can be submerged in the water and still survive. It absorbs oxygen dissolved in the water through its skin.

Other amphibians also extract oxygen and release carbon dioxide through their skin. They can do this because the skin is filled with blood vessels to carry the gases. The skin is not as strong a barrier to the transfer of gases as it is in other vertebrates.

For instance, a salamander is a type of amphibian. Some adult salamanders, such as hellbenders, do not have gills or lungs. The hellbender is the largest salamander found in the United States. It grows to a length of about 2.5 feet. Despite its large size, it has no lungs but instead takes in oxygen directly through its skin. Deep wrinkles increase the surface area exposed to the atmosphere. The wrinkles increase the amount of oxygen it can absorb.

Mud puppies, another species of salamander, spend most of the day on the river or lake bottom. Although lungs develop, mud puppies depend on their gills to extract oxygen from the water throughout their lives. Mud puppies can grow

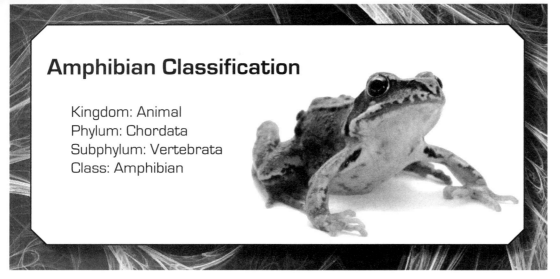

Amphibian Classification

Kingdom: Animal
Phylum: Chordata
Subphylum: Vertebrata
Class: Amphibian

to a length of one foot.

The skin of salamanders, frogs, and toads does not provide the degree of protection as the skin of other vertebrates. Fish and reptiles have scales. Birds have feathers. Mammals have fur, wool, or hair. Amphibian skin has none of these structures. Instead, amphibian skin is usually soft and moist. It has no scales.

Because they dry out so easily, amphibians avoid direct sunlight and are more active at night. Most spend at least part of their lives in moist surroundings.

Like fish, they are cold-blooded. Amphibians do not have set body temperatures. Instead, their body temperatures vary with the environment. Amphibians cannot survive cold climates. Antarctica, the continent at the South Pole, has no amphibians. The largest number of species of amphibians lives in the tropics.

The total number of amphibians has declined steadily since counting them began in the 1980s. Most biologists believe a combination of factors has caused the decrease in amphibians. These factors include water pollution, climate change, and acid rain — rain that has absorbed sulfur, a chemical in air pollution. Another factor is the loss of wetlands. Farming and urban growth destroy wetlands. Amphibians are forced into smaller areas. Amphibians in close contact with one another catch disease, fungus infection, and parasites more easily from one another.

Amphibians have a complex life cycle. Part of the time they live in water, part of the time on land, and they return to water to lay eggs. Such a complex life cycle can reveal the condition of the environment. Changes in water, land, and air have a profound effect upon amphibians such as frogs. A sudden drop in the number of frogs can be an early warning that the environment is changing in a way that could also be a risk to humans.

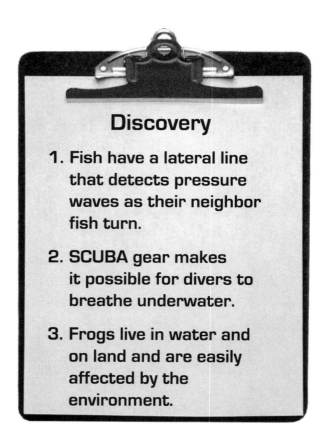

Discovery

1. Fish have a lateral line that detects pressure waves as their neighbor fish turn.

2. SCUBA gear makes it possible for divers to breathe underwater.

3. Frogs live in water and on land and are easily affected by the environment.

Questions

T F   1. Biologists believe that new plant and animal species are unlikely to be discovered.

A B   2. The one with the greater number of animal species is (A. invertebrates   B. vertebrates).

A B C D   3. The one that is a vertebrate is (A. coral   B. fish   C. lancet   D. snail).

A B C D   4. The classification of vertebrate is a (A. class   B. kingdom   C. phylum   D. subphylum).

A B   5. Fish are (A. cold blooded   B. warm blooded).

6. What special sense organ makes it possible for a school of fish to turn together? _____

A B   7. Jacques Cousteau wrote (A. Twenty Thousand Leagues Under the Sea   B. The Silent World).

A B C D   8. Amphibians include frogs, toads, and (A. catfish   B. goldfish   C. salamanders   D. salmon).

9. Amphibians can breathe through gills, lungs, and _____.

T F   10. Rather than feathers, wool, or hair, amphibian skin is protected by scales.

A B   11. Amphibians are (A. cold blooded   B. warm blooded).

A B   12. Biologists believe the number of amphibians is on the (A. rise   B. decline).

Explore More:

What is the deepest part of the ocean? Have explorers managed to dive to that depth? Did they find life there? Choose a hero of undersea exploration and describe his or her discoveries. What are the latest efforts to develop gill backpacks (also called artificial gills) for human underwater swimmers?

Some fish, such as salmon, travel from saltwater to freshwater to spawn. Why do biologists think they make the difficult trip upstream to lay their eggs?

Some people have an aquarium for keeping fish. What must be done to maintain healthy fish in an aquarium? Do you or your friends keep amphibians, reptiles, birds, and mammals as pets?

Fishing is a pleasant hobby for many. What are the most popular game fish? Fishing is also a commercial way of earning a living. What fish are sources of food?

Amphibians include frogs and toads. How do frogs and toads differ from one another?

Chapter 11

Reptiles

Reptiles are vertebrates. They have a backbone. Reptiles include snakes, lizards, turtles, crocodiles, and alligators. They have scales on their bodies and except for snakes, have four legs with feet that end in claws. Reptiles breathe air with lungs and most reproduce by laying eggs.

Reptiles are cold-blooded. A cold-blooded animal is one that has no internal method to keep its body temperature constant. Chemical reactions that power muscles require heat. For that reason, on cool mornings, reptiles make an effort to get warm so they can be active and run after prey. Without heat, they are slow and sluggish. Turtles sun themselves on logs floating in a pond. Lizards and snakes use warm rocks and sunlight to raise their body temperatures.

Explore

1. Why must reptiles warm their bodies before seeking prey?
2. How do some snakes see prey in the dark?
3. How do geckos walk upside down on a ceiling?

Reptile Classification

Kingdom: Animal
Phylum: Chordata
Subphylum: Vertebrata
Class: Reptile

Collared Lizard

Chinese Box Turtle

Columbian Boa

Once fully warm, they can dart about faster than their warm-blooded prey, such as mice.

Reptiles do succeed better in hot climates than in cold climates. Reptiles such as snakes and lizards are abundant in the southwestern United States and northern Mexico. They control their body temperature by first darting into the sun to absorb heat, then sprinting into shade to keep from getting too warm. Reptiles have no sweat glands and cannot cool themselves. When the temperature becomes too warm, reptiles will seek shade or hide under a rock. Some go into a sort of inactivity called torpor.

Unlike fish and amphibians, reptiles are land creatures. Although some reptiles live in water, they must come to the surface to breathe air and crawl onto the land to lay eggs. Even ocean-going reptiles such as leatherback sea turtles return to the land to lay their eggs in sand along the beach.

Snakes and lizards usually lay their eggs in places that provide natural cover, such as spaces under rocks or hidden openings under fallen trees. Turtles usually bury their eggs in sandy soil. Most reptiles abandon the eggs once they are laid. They provide no care for the young after they hatch.

However, female crocodiles and alligators make a nest for their eggs. They wallow out depressions along the banks of rivers and heap together dead limbs and leaves to fill in the holes. Female crocodiles remain near the nests and chase off animals that might eat the eggs.

Several species of snakes do not actually lay eggs. Instead, the female keeps the eggs in her

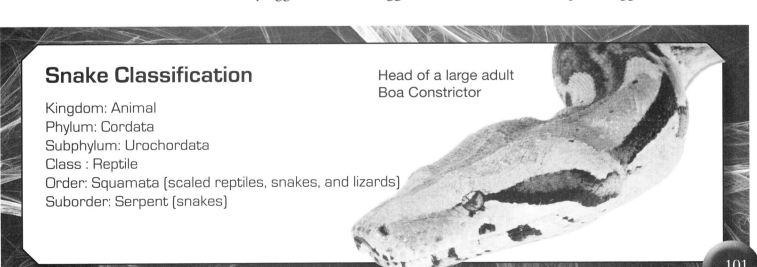

Snake Classification

Kingdom: Animal
Phylum: Cordata
Subphylum: Urochordata
Class : Reptile
Order: Squamata (scaled reptiles, snakes, and lizards)
Suborder: Serpent (snakes)

Head of a large adult Boa Constrictor

Snakebite!

A bite from a snake without fangs comes only from the snake's teeth. It does not inject poison. It inflects a series of scratches or tiny punctures. A bite from a non-venomous snake is unlikely to be a problem, but any bite should be examined by a medical doctor to ensure that it is, in fact, not from a venomous snake. Like any other breaks in the skin, the tiny punctures offer a site for bacteria to enter the body. They should be treated with an antiseptic to prevent infection.

Snakebites by venomous snakes are more serious but few in number. Often, bites are not by snakes in the wild. Instead, they are by snakes kept as pets, those in medical research laboratories, or in reptile farms. In the United States, venomous snakes bite about 1,000 people each year, and about 15 people die from bites. This is far less than the deaths caused by bee stings. Even mosquitoes produce more deaths by diseases that they carry.

What causes the venom of a snake to be toxic to humans? Venom is an enzyme. Normally, enzymes in the human body promote helpful chemical reactions. In the case of snake venom enzyme, it causes chemical reactions that destroy protein. It injures nerves, ruptures blood vessels, or dissolves flesh.

The best treatment for snakebite is not to get bitten in the first place. To avoid snakebites, avoid putting your hands in dark openings in rocks where you cannot see. When hiking, don't step over a log without first looking at what might be on the other side.

What should be done to treat snakebite? Almost every remedy that was used in the past has now been discredited. Experts caution against cutting the wound to drain or suck out toxin. Deep incisions can sever a nerve, cut a tendon, slice into a blood vessel, and cause loss of blood. Infection is more likely when the skin is cut, and there is no evidence that a lot of venom can be removed. Using a strong tourniquet to stop the spread of venom is also dangerous. A tourniquet squeezes blood vessels closed. It can damage tissues and cause cells to die because blood flow is cut off.

The venom of some snakes reduces the amount of oxygen that the blood can carry. For that reason, keep the victim calm without undue exertion. Then rapidly transport the victim to an emergency medical facility. Quick medical aid is the best treatment. Most hospitals have antivenin for poisonous snakes found in the United States. Antivenin is an enzyme that neutralizes snake venom and reduces its harmful effects.

body until the eggs hatch. The young snakes are fully developed and ready to fend for themselves.

Snakes belong to the class Reptiles. In the classification system, the next division is order. In the case of snakes, they are of the order Squamata (SKWAY-ma-tah, scaled reptiles), and a member of the suborder Serpent.

Snakes have no arms or legs. Because of their tapering body, most snakes have only one functioning lung. They have no room for two full-size lungs. They have no vocal cords and no external ears. Snakes have no eyelids, so their eyes are always open, even in a blowing sandstorm. In addition, their eyes do not move. They have to move their heads to look about.

Imagine their condition: no arms, no legs, only one lung, no vocal cords, and no eyelids. Yet, snakes survive and thrive despite these disabilities.

Snakes have a red, forked tongue that they flick about in a way that looks sinister. However, they flick it in and out of their mouths to sample airborne chemicals the tongue collects. The snake brings its tongue back to a special organ at the roof of the mouth. Nerves in the organ interpret the nature of the chemicals and send the information to the brain. The tongue helps identify whether an animal is nearby that might serve as a meal.

In addition to a tongue that detects prey, some snakes have built-in night vision sensors. They can see in total darkness by detecting heat given off by their prey. For instance, a copperhead snake has a pit about one quarter inch deep located between the eye and the nostril on each side of the head. Unlike eyes that can see only visible light, the pits detect infrared radiation, or heat rays. A snake equipped with such an organ is called a pit viper. In total darkness, a pit viper can detect an animal's body heat and identify its position and distance. A motionless mouse in a dark room is no match for the copperhead. The snake can accurately aim its strike to take the mouse.

A snake can swallow about anything it can kill because its jaws are not permanently attached at the back. They can separate from one another. A snake can open its mouth really wide.

Venomous snakes, such as the copperhead, have two needlelike teeth, or fangs. The fangs have a hole down the middle that delivers the venom. The fangs connect by tubes to two venom glands on each side of the head. Most snakes coil to get ready to strike. Then the head and part of the body uncoils suddenly. The snake arrives at its target with its mouth wide open and the fangs ready to plunge into its victim.

The strike of the fangs is more like a stab than a bite. Some snakes have fangs so long they can kill

Copperhead

Cottonmouth

without venom. The prey dies as if stabbed with two sharp knives. In addition to killing prey, a venomous snake strikes as a defense against animals or humans that it perceives as a threat. Four venomous snakes are found in the United States: copperheads, moccasins, rattlesnakes, and coral snakes.

Copperheads are the most common venomous snakes of the eastern United States. Copperheads have copper-colored heads, and the rest of their bodies are tan or pinkish. Most range in length from two to three feet. They eat mice, other snakes, birds, frogs, lizards, and insects.

Like all reptiles, copperheads reproduce by eggs. However, the female keeps the eggs in her body. The young hatch inside the mother's body, and she gives birth to live young.

Sometimes copperheads are found in pairs. A person may see the snake, jump away from it, and then land next to the other one. Copperheads bite more people than any other venomous snake in the United States. The copperhead injects venom that destroys the blood's ability to carry oxygen. It also attacks the linings of blood vessels and causes them to break. The area around the wound swells, becomes discolored with blood, and the skin breaks. However, copperhead venom is not as toxic as that from other venomous snakes in the United States, so a bite seldom proves fatal.

A moccasin snake is also known as a cottonmouth or water moccasin. The cottonmouth is aquatic. It lives in and around water. The cottonmouth is found in swampy areas in the southeastern states and west to Texas. The snake grows to a length of three to four feet. On land, it eats reptiles and small mammals. Some people think a snake must be on land to strike. That is not the case. The cottonmouth also feeds on fish, frogs, and turtles that it kills in the water.

Like the copperhead, it is a pit viper with a highly venomous bite that destroys red blood cells. Most snakes hide when alarmed, but a moccasin stands its ground when threatened. It holds its jaws wide open to reveal its fangs and the cottony white interior of its mouth. The aggressive display

Cobra

A cobra is a type of very venomous snake that lives mainly in warm, tropical climates in Asia, especially India, and in Africa. Cobras are noted for the way they flare their necks when angry. The name *cobra* is a Portuguese word meaning "snake with a hood." Ribs expand outward and stretch the skin behind the cobra's head to make the hood.

A cobra subdues its prey — lizards, fish, frogs, and other snakes — with its venom and swallows them whole after they die. The venom contains a powerful chemical that affects the nervous system. The toxin paralyzes muscles. Mammals breathe with a diaphragm that moves air in and out of the lungs. Cobra venom paralyzes the muscles that operate the diaphragm. The lungs stop working. Breathing slows down. The victim suffocates.

A cobra can deliver with one strike about eight times as much venom as it takes to kill a human being. In the United States, the only snake as deadly is the eastern diamondback rattlesnake. However, adult cobras can control how much venom they release and seldom use the full dose. Even so, about one person in ten attacked by a cobra dies from the strike. Cobras are a real danger in India, where people are poor and often walk barefoot through fields.

Cobras have hypnotic-looking, bronze-colored eyes. People who come into contact with a cobra report that it is difficult to look away from it. Because of its unusual appearance, snake charmers work with cobras. They play a flute and the cobra appears to sway to the music. Actually, the motion is because the snake charmer waves his flute back and forth. The cobra's head follows the end of the flute.

The three best-known types of cobras are the common cobra, spitting cobra, and king cobra. Snake charmers use the common cobra. It is about six feet long and has a large, brightly colored hood with an unusual eyeglass-shaped pattern on its head.

Most cobras coil to strike. But spitting cobras can shoot their venom a distance of five to ten feet through the air with enough accuracy to temporarily blind a person.

The king cobra is the largest type of cobra and the world's largest venomous snake. Adults reach a length of 12 to 18 feet.

is a defense measure to ward off attack. A moccasin usually bites a human only when stepped upon or picked up. Its bite can be deadly and is certainly painful, even if the victim survives.

The rattlesnake is a well-known venomous snake in the United States. It is a patient hunter, lying in wait for hours or even days along a path waiting for small animals such as mice, rats, or chipmunks to get within striking distance.

The best-known feature of the rattlesnake is its rattler at the end of its tail. The sound is more of a buzz than a rattle. Many snakes shake their tails when disturbed. If the tail strikes dry leaves or twigs, it makes a rustling sound. Rattlesnakes do not

Western diamondback rattlesnake

need to strike other objects but merely coil up and shake their tails.

The rattle, however, is only used when it wants its presence to be known. If a rattlesnake is hunting in the same field as cattle or horses, it buzzes them when they approach too closely. The buzz says, "Watch where you are walking and don't step on me!" The unmistakable rattle is an early warning device to alert other animals or humans that it feels threatened.

The rattle is a series of nested, hollow beads. A rattlesnake gains a new one each time it sheds its skin. A larger number of rattles is the sign of an older snake, but the exact age in years is not given by the number of rattlers. If a rattlesnake is well fed and grows rapidly, it may need to shed its skin three or four times a year. In addition, if the rattler grows too big, it may catch in tight places and be broken off.

Because of its fierce reputation, most people exercise caution when they see a rattlesnake and are seldom bitten. Those that do get bitten often fail to retreat and underestimate how far rattlesnakes can strike.

Rattlesnakes can strike from a farther distance than at first seems possible. When coiled, they may appear to be only a few feet long. People incorrectly estimate the striking distance at one-third of the animal's body length. But rattlesnakes can be eight feet long and can strike two-thirds of their length. The strike is blindingly fast, far faster than the eye can follow. Their heads become merely a blur. Rattlesnakes can control the amount of venom that they use. For defense, they may inject none at all. But a frightened snake, or one that has become injured (because someone stepped on it) may not be able to use such self-control.

The venom of a rattlesnake is double acting. It destroys blood vessels in the same way as the venom of copperheads. In addition, the rattlesnake produces a chemical that affects the nervous system. It enters the bloodstream and slows the action of the heart and lungs. Prey lose endurance to struggle and escape.

Diamondback rattlesnakes are the largest and most lethal of the rattlesnakes. Some reach eight feet long, and because they have thick bodies, they can weigh 20 pounds. Some diamondbacks are cooked and eaten as food.

The coral snake is the fourth venomous snake in the United States. Coral snakes are strikingly beautiful with glossy red, yellow, and black colored bands along their bodies. One species is found in the southeastern United States. Another species lives in the deserts of Arizona and New Mexico. Both have tiny eyes in heads that are the same width as their slender bodies.

The venom is deadly, but it is not fast acting. Coral snakes must hold onto the victim until the venom does its work. Other than their distinctive color, coral snakes have no other way to warn of their presence. They tend to stay out of sight under leaves. They come out at night to hunt smaller snakes and lizards.

Their beautiful color and small size hides the

Lizard Classification

Kingdom: Animal
Phylum: Chordata
Subphylum: Urochordata
Class : Reptile
Order: Squamata (scaled reptiles)
Suborder: Lacertilia

fact that coral snakes carry powerful venom that can kill a person. The toxin of a coral snake does not attack blood vessels. Instead, it damages the nervous system. Sometimes a person who has been bitten feels no pain because the venom numbs the nerves. In the most severe case, the body becomes paralyzed and the person is unable to breathe.

Venomous snakes found elsewhere in the world have venom similar to that of the four types of venomous snakes (copperhead, moccasin, rattlesnake, and coral snake) in the United States. With a length of 18 feet, the king cobra is the largest venomous snake in the world. Its venom, similar to that of the small coral snake, attacks the nervous system. Muscles stop working, the heart slows, and breathing ceases.

The longest snake in the world is the python at a record length of more than 30 feet. The largest snake by weight is the anaconda (an-uh-KON-duh) of South America. It can be as long as 20 feet and weigh more than 200 pounds. Both the anaconda and python are constrictors. They kill by squeezing their victims so hard they cannot breathe. A python can swallow an animal such as a newly born water buffalo that may weigh 150 pounds. In the United States, the most common constrictor is the rat snake. It feeds on mice and rats.

Both snakes and lizards are placed in the same order because they have a very similar design. Like snakes, lizards are cold blooded and have dry, scaly skin. Snakes are in the suborder Serpent. Biologists put lizards in the suborder Lacertilia (LAS-uhr-TIL-uh). The word *Lacertilia* is merely the Latin name for lizards.

Most lizards have four legs with feet that end in claws. Unlike snakes, lizards have external ears and movable eyelids. Lizards have slender bodies and long tails. The tail can be a method of defense. If caught by the tail, lizards can give it up and then grow a new one.

Lizards, like all reptiles, are cold-blooded. Lizards must warm to 80ºF

Coral snake

before they can begin their day's activities. Because of their need for warmth, lizards are most active during the day. On a cold morning they will lie against a rock or tree trunk that is also absorbing sunlight. After they become warm, they are nimble and dart about quickly.

Lizards that live in deserts often can survive without drinking water. They extract moisture from leaves and the small insects or animals they eat. Digesting food and burning it for energy generates water as a by-product, too.

Only one lizard — the Gila monster — is poisonous, and it lives in the deserts of the southwestern United States. The body of a Gila monster is covered with orange and black beadlike scales. It can reach a length of about two feet. The Gila monster's mouth clamps onto its prey and releases the venom as it chews. It is not a particularly fast lizard and seldom poses a threat to humans. No human fatalities have ever been recorded. However, being bitten by a Gila monster is not a pleasant experience.

The chameleon is a lizard that has the ability to change skin color. When the chameleon is frightened, it releases hormones that cause the color change. Many people believe the chameleon changes its color to match that of the background. But biologists are not so sure. They think that in a time of stress it becomes frightened and releases an enzyme that triggers the color change. If it were visible before, the new color is likely to make it less visible and more likely to be the same color as the background.

Lizards feed on a wide variety of food. Some dine on insects, others on vegetables. Some find bird eggs a nice snack. Others are not very picky eaters and will even eat an animal that they find already dead. Unlike snakes, lizards cannot unhinge their jaws, so they must limit what they eat to something more their size.

The largest lizard of all is the Komodo dragon. It lives on Komodo Island and nearby islands in Indonesia. The Komodo dragon can reach a length of 10 feet and weigh more than 300 pounds. It has strong legs and a powerful tail. Its large size means it can target larger prey, including deer.

Lizard species have a tremendous variety in color, appearance, and size. The gecko is a small lizard found mostly in the tropics. The gecko is a noisy little lizard. It is named after the sound it makes, a dry, clicking "gecko" sound.

The wonder of the gecko is not the sound it makes, but its ability to climb smooth, vertical walls. How it is able to do this is still not fully understood. A gecko has feet pads with millions of fine, hairlike projections. Scientists believe the hairs are so thin they interact with the molecules of the surface the gecko is walking across. Even the smoothest glass is no match for the gecko. It can climb the side of a window and run upside-down across a smooth ceiling. People tolerate geckos as houseguests because they eat insects.

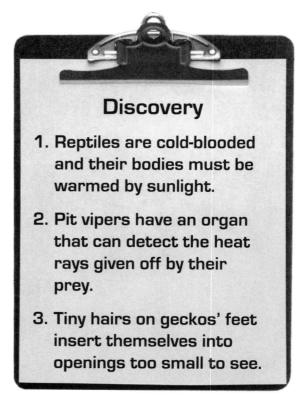

Discovery

1. **Reptiles are cold-blooded and their bodies must be warmed by sunlight.**

2. **Pit vipers have an organ that can detect the heat rays given off by their prey.**

3. **Tiny hairs on geckos' feet insert themselves into openings too small to see.**

Questions

T F   1. Reptiles are vertebrates.

A B   2. Snakes are (A. cold-blooded B. warm-blooded).

A B C D   3. The one that is not a reptile is a (A. frog   B. lizard   C. snake   D. turtle).

T F   4. Most reptiles have sweat glands to cool themselves.

A B C D   5. The reptile that protects her nest is the (A. crocodile   B. Gila monster   C. sea turtle D. snake).

       6. Why do snakes flick their forked tongues in and out? _____

A B   7. The pit of a copperhead is an organ to sense (A. heat rays   B. odors).

A B C D   8. The recommended treatment for a snake bite is to (A. cut the wound to release venom   B. keep the victim calm and transport him to the hospital   C. use a tight tourniquet to stop the spread of venom   D. use ice to reduce swelling).

A B   9. The one that is more likely to be found in water is the (A. coral snake   B. moccasin).

A B C D 10. The toxin of a coral snake (A. attacks blood vessels   B. blinds a person's vision   C. causes deep puncture wounds   D. damages the nervous system).

A B   11. A python attacks by (A. spitting poison   B. squeezing its victims).

A B   12. Lizards are (A. cold blooded   B. warm blooded).

A B C D 13. The poisonous lizard is the (A. chameleon   B. gecko   C. Gila monster D. Komodo dragon).

       14. The lizard that can change color is the _____.

       15. Matching:

           __ The most common venomous snake found in the eastern United States   a) anaconda

           __ Is cottony white inside its mouth   b) copperhead

           __ Can weigh 20 pounds and is cooked as food   c) coral snake

           __ Has glossy red, yellow, and black bands along its body   d) diamondback rattlesnake

           __ Is the largest venomous snake   e) king cobra

           __ Is a constrictor and is the largest snake by weight   f) moccasin

EXPLORE MORE:

The different species of reptiles are vast, and many more exist than have been described in this chapter. Choose one or two of interest from the list below and explore more about them. Describe the type of environment in which they make their homes, their food, how they lay their eggs, and other interesting facts.

     Terrible Lizards (dinosaurs)
     Rat snake
     "Flying" snakes
     Iguanas
     Turtles
     Crocodiles

Chapter 12

Birds

Birds are vertebrates. In addition to backbones, they share some features with other vertebrates. They walk on two legs, as do primates like humans. They lay eggs, as do fish, reptiles, and amphibians. Birds are warm-blooded, a feature they share with mammals. Like other warm-blooded animals, such as horses and rabbits, birds have the ability to keep a constant body temperature.

Their design differs from other animals in many ways, too. Bird eggs are hard-shelled, while those of reptiles and amphibians are flexible. Parent birds can sit on a nest without breaking the eggs. Birds have a body temperature that is warmer than most other warm-blooded animals. The natural internal temperature of birds ranges between 104°F and 108°F, depending on the species. The set body temperature of birds is six or

Clipboard Explore

1. How can the same species of bird have different appearances?

2. Why did the dodo and passenger pigeon become extinct?

3. How does migration reduce the number of predators?

Bird Classification

Kingdom: Animal
Phylum: Chordata
Subphylum: Vertebrata
Class: Aves (birds)

ten Fahrenheit degrees higher than that of humans, 98.6°F.

Birds live at a fast pace to maintain their high body temperature. Their hearts beat about five times as rapidly as the human heart, and birds breathe more quickly, too. A bird in flight breathes seven times per second. Its heart pumps blood at about ten beats per second. Compare respiration and heart rate of birds with humans. Even when running, a human breathes about once per second, and the heart seldom beats more than three times a second.

Birds have feathers, a feature they share with no other animal. For most birds, feathers cover their entire body, except the legs and beak. Feathers aid in insulating birds from cold weather. Ducks and geese have feathers with oil that sheds water, so they can swim in water without getting soaking wet.

Throughout history, birds have served as an important source of food for humans. We may eat eggs for breakfast, feast on turkey on thanksgiving, and enjoy fried chicken throughout the year. Other birds that serve as food include duck, geese, quail, and pheasants. Birds grown for human consumption are referred to as poultry.

People also keep birds as pets. They keep canaries and parakeets for their songs and parrots for their ability to mimic human speech. Humans enjoy the showy display of the male peacock, the bright plumage of the scarlet macaw, and the beautiful color of the green parrot.

Some birds work with humans. During medieval times, an opposing army would put a castle under siege. No one could get in or out. However, falcons were trained to catch rabbits and other small animals and fly them back to the castle. By growing vegetables inside the castle walls, those trapped inside could survive for months on rabbit stew. The Bible tells about Elijah, who was fed by ravens. Read 1 Kings 17:1–6.

Birds also figure in the Bible in the story of Noah's ark. The birds were used as scouts to discover if dry land had once again appeared as the water receded. After releasing a raven, Noah sent out a dove. The dove returned because it could not find dry land. Seven days later, he sent out a dove and it returned with a fresh olive leaf. Then a week later, another dove did not return, revealing it had found dry land (Gen. 8:7–12.)

Before the telegraph and telephone, people used pigeons for quick, long-distance communication. Trainers transported the birds away from their home roost and released them. The birds naturally flew back to the place where they received food and shelter. Over time, trainers released the pigeons farther and farther away from home. Birds trained to carry messages became known as carrier pigeons.

During the wars with Napoleon, the British dispatched carrier pigeons back to England to report on the progress of battles. The Duke of Wellington finally defeated Napoleon at Waterloo, Belgium, on June 8, 1815. The British learned of the victory first by a carrier pigeon. Observers at Waterloo wrote the information on a very thin piece of paper and rolled it into a small message tube attached to the bird's leg. The pigeon flew from Waterloo across the

Homing pigeons in flight

English Channel to London. It arrived four days before the same message carried by horse and ship told of the victory.

In Africa a type of bird, known as the honey guide, points tribesmen to wild honey. The bird flies around and chirps, calling for natives to follow. Tribesmen follow the bird. The honey guide will land on a tree limb and chirp insistently, waiting for the humans to catch up. Then it flies ahead impatiently. Finally it comes to rest on the tree with honey. Natives make a fire and smoke the hive, pulling out the rich and sweet honeycomb of the wild honeybees. They leave some of the wax and larvae for the honey guide.

Bird watching is a pleasant pastime for millions of people. Bird watching is one of the most popular hobbies worldwide. It became popular when people realized they could identify birds without shooting them. A hundred years ago, people went bird watching with a shotgun. They would kill the bird and bring it back to a museum or nature center. There they compared the dead bird with technical descriptions. Books identified birds by biological features and called for measurements of the beak, talons, and wings. Sometimes the book described features found only in internal organs. This information could not be seen from a distance.

Roger Tory Peterson (1908–1996, American naturalist) made bird watching popular because he found ways to identify birds from a distance. Bird books listed all of a bird's features, even those that did not aid in telling one bird from another. For instance, raptors have talons shaped for catching and grasping prey. Once a bird was identified as a raptor, showing its talons was not much help, because all talons looked the same from a distance.

Roger Tory Peterson realized he could identify almost every species by a few visible clues. He looked for markings that made one bird different from others. For instance, in the eastern United States, the American goldfinch is the only small yellow bird with black wings. The screech owl is the only owl with ear tufts. Most books with illustrations of birds were large and had to be left at home. Yet Roger could replace a huge book with a single page of his drawings. Roger was studying painting in New York City. In six months he identified 100 different species in the parks of New York City.

Bird-watching friends urged him to publish his book. The year was 1934, in the depths of the Great Depression, a time of unemployment and hard economic times. A book with illustrations cost far more to produce than one with straight text. Publishers estimated the book would have to sell for $2.75. This made it a costly book when you consider that in those days restaurants served full meals for only 35 cents. Who could afford such an expensive book about bird watching? Four publishers turned him down.

The publishers did not realize that people were looking for cheap entertainment. A book on bird watching would bring enjoyment for days, months, or even years. Finally, in 1934 Houghton Mifflin in Boston issued Peterson's *Field Guide to Eastern Birds*. They cautiously printed only 2,000 copies. All copies sold in a couple of weeks, and the publisher had to go back and print more.

As the years passed, Roger wrote and illustrated guides about other subjects, but birds remained his favorite subject. Once, at an outdoor gathering of bird watchers, he sat quietly with his eyes closed. Others walked the trails to see what birds they could identify. They came back and described all of the birds they had found. Roger held up a sheet of paper. On it he had listed the birds that he identified by the bird songs and calls alone. His list was longer than the one made by the group on the trails.

Roger Tory Peterson died in 1996 at the age of 87. He received many honors during his lifetime, including a bird named after him. The cinnamon screech owl lives in the misty forests high in the mountains of Ecuador. This beautiful little owl was discovered in 1976 and given the scientific name *Otus petersoni*.

Bird watching can be a never-ending hobby because the number of different species of birds is so vast. Carl Linnaeus made the first scientific list of bird species. In 1758 he listed 564 species. Today ornithologists (or-nuh-THOL-uh-gistz) — bird experts — estimate that about 8,500 to 10,000 bird species exist worldwide. A bird watcher who manages to identify 5,000 different species is considered at the top of his or her hobby.

A listing of bird species does change. As the earth is explored, biologists discover new species. Birds are members of the same species if they can reproduce and produce young that are also capable of breeding. The number of species can decrease because ornithologists discover birds that vary in size and color may actually be of the same species. Birds can look different depending upon whether they are male or female, juvenile, or whether they have winter or summer plumage. A bird seen in winter may look considerably different than the same one seen in summer. At first, biologists may think it is a different species.

In addition, the number of bird species is reduced when birds become extinct. Perhaps as many as 500 bird species have become extinct since the year 1500. As one example, in the late 1500s, Portuguese sailors landed on Mauritian, an island 500 miles east of Africa in the Indian Ocean. They discovered a previously unknown bird. It had not seen humans before. It greeted the new visitors with curiosity and walked up to them. It showed no fear.

The plump bird weighed more than a turkey and had a large, hooked beak. Its stubby wings were useless for flight. Because of its thick, heavyset body and inability to fly, it appeared clumsy and stupid. The sailors called it a dodo, from the Portuguese word *doudo*, meaning "silly." The *do* in dodo is pronounced as "dough" in doughnut.

Fresh meat was a luxury for sailors, so they were happy the animal could be so easily killed, cooked, and eaten.

A few years later, Dutch farmers settled the island of Mauritian where the dodo lived. Because the bird did not fly, it had to lay its eggs on the ground. The eggs became a tasty meal for pigs that escaped from farmers' barnyards. Wild dogs took their toll, too. In only 80 years, the last of the clumsy-looking birds had died out.

A fascinating event occurred after the dodo was lost to the world. A type of tree on the island began dying. Its seeds did not sprout. Naturalists investigated and found that the tree produced a seed with a hard shell. The covering prevented the seed from sprouting. The dodo ate the seed, and

John J. Audubon and the Birds of America

John James Audubon grew up in France where he developed an interest in nature and especially drawing birds. As the dark clouds of Napoleon's wars developed in Europe, Audubon's father sent the young man to an estate he owned in America. A hard-working Quaker family ran the farm in Pennsylvania. Eighteen-year-old John Audubon soon filled his room with collections of birds' nests, feathers, minerals, and flowers.

One day he noticed some phoebes nesting in a cave. People claimed that the birds returned to the same place each year. He decided to test the idea. He sat silently and read until the birds became used to his presence. Each day he moved closer. Finally, he could touch the birds and they did not become alarmed. He tied silver threads to their legs. The next year, his birds with the silver threads returned to the same cave.

The Bakewell family lived next door. There he met Lucy Bakewell. Within a year, John and Lucy decided to marry. Lucy's father refused. He questioned John Audubon's ability to earn a proper living for his daughter.

To prove himself a provider, Audubon purchased supplies and headed west to the frontier. He opened a general store in Henderson, Kentucky. It earned a small but steady income. After five years of waiting, he and Lucy finally married.

For the next ten years, he ran the store but escaped into the wild countryside as often as possible. As Audubon sought out new birds to draw, his general store business suffered. Lucy realized that if her husband were to succeed, it would be as an artist. She told him, "Give yourself full time to drawing birds."

Most artists purchased stuffed birds and drew them in the studio. Audubon sketched the birds while they were alive and moving. Back at the studio, he wired dead birds to give them the lifelike poses he had seen in the wild. Although he worked primarily with watercolor, he used whatever would give the image the right texture and color. His paintings showed the shimmering brilliance of feathers.

He took his collection of bird drawings to Philadelphia, the center of publishing in the United States at that time. George Ord, Philadelphia's best-known naturalist, inspected the drawings. He dismissed Audubon's work. He told everyone Audubon's images were scientifically inaccurate.

Audubon could not imagine why Ord was so negative. Later, the truth came out. George Ord was working on a book about birds. Sales of Ord's book would suffer if Audubon succeeded in finding a publisher. George Ord spread the false story to earn more money. The damage had been done. American publishers refused to publish Audubon's book.

At a friend's suggestion, Audubon set sail for Liverpool, England. People flocked to see his paintings. He made 500 dollars simply by putting the paintings on display and charging admission.

He decided to select his best bird pictures to be engraved and hand colored to make a book. He drew birds at their actual size. The pages had to be two and a half feet wide by more

Painting of Carolina parakeets, from J.J. Audubon's *Birds of America*, 1829.

than three feet high. The book would be frightfully expensive — a thousand dollars! He decided to pre-sell copies to raise money for the cost of printing. With endless patience, he showed his paintings and encouraged well-to-do people to buy the book. He assured success when he persuaded George VI, King of England, to purchase a copy.

Audubon could not stay away from home any longer. After a warm family reunion, he collected his best paintings to take back to London. In June 1838 the last print was finished. It had taken 12 years. He called his book *The Birds of America*. His book displayed 1,065 life-size birds in full color. John Audubon's bird drawings succeeded as both art and science.

Audubon painted a bird that later became extinct. The brightly colored Carolina parakeet had a green body and yellow and orange head. It had beautiful tail feathers. The bird was killed for its feathers to adorn women's hats. Farmers killed them because they ate the fruit in their orchards. The last known Carolina parakeet died in captivity in 1918.

its digestive process removed the covering. Without the dodo to eat and spread the seeds, the tree disappeared, too.

The United States has lost its share of birds. In the early 1800s, residents of Henderson, Kentucky, were on the flight path of migrating passenger pigeons. One day, the mid-day sunlight became dusky as if from an eclipse. The people stared up in astonishment as a mile-wide stream of pigeons passed overhead. For three hours the migration continued. The single flock contained more than a billion (1,000,000,000) birds.

No laws protected birds. Their numbers appeared endless. Shooting birds for food was common. Pioneer families depended on the birds for food and hardly made an impact on their numbers.

This changed with the coming of the railroads. Rapid transportation allowed freshly killed birds to be shipped to restaurants and meat shops along the East Coast. Passenger pigeons were large birds, about a foot long, and restaurants served them. Professional pigeon hunters slaughtered the birds by the millions. In the 1870s people noticed that the number of passenger pigeons was in deep decline.

Frantically, naturalists tried to breed the birds in captivity. Unable to migrate to their spring nesting grounds in Canada, the birds failed to reproduce. No longer did their vast flocks darken the sky. As the 1800s ended, not a single bird was seen in the wild. The last passenger pigeon died in the Cincinnati Zoo in 1914.

Perhaps if the dodo, passenger pigeon, and others had survived, scientists might have discovered important uses for them. Even without a specific use to benefit people, a particular species of birds may be essential for the well being of other life. They make the earth a richer and more interesting place.

Most birds are designed for flight. They must be lightweight, streamlined, able to control their flight, and stay warm in cold air. Feathers help them do this. Feathers are lightweight and give birds a streamlined shape to reduce air friction. Feathers at the tips of the wings and those in the tail serve for balancing and steering. Feathers are good insulators and keep the birds warm as they fly through cold air. Water birds have feathers coated with oil that sheds water. When swimming or diving, feathers keep the birds dry.

The color of feathers often differs between the male and female birds. Usually, males have brighter and more distinctive plumage. Females tend to be much less brightly colored. Their feathers are often brown or gray, the same color as their nests. Females with their drab colors are harder for predators to see. Should the nest be threatened, the more brightly colored male will distract the predator and lead it away.

Each type of bird has a beak that is designed for the food it eats.

Vuture   Hawk   Woodpecker   Pelican

Once in flight, birds must provide their muscles with plenty of oxygen. Birds ensure a good supply of oxygen with an unusual respiratory system. When birds inhale, part of the fresh air goes into the lungs, but part of it flows into air sacs located throughout the body. When birds exhale, used air flows out of the lungs and is replaced by the air from the air sacs. Whether breathing in or out, a bird's lungs receive a constant supply of fresh oxygen.

Birds must also provide their muscles with plenty of high-energy food. Unlike cattle and horses that can survive on grasses, birds eat foods high in energy, such as seeds and insects.

Each type of bird has a beak that is designed for the food that it eats. The sparrow has a small but powerful beak for cracking the outer husk of seeds. The woodpecker has a strong, pointed beak for drilling into wood to find insects. The hummingbird has a long, thin beak that is like a hollow straw. The hummingbird can sip nectar from flowers. Owls and hawks have sharp, hooked beaks for ripping flesh from the animals they catch. The pelican has a long, straight beak with a pouch of skin below it for holding fish.

Some birds have a long, sharp beak for spearing fish. For instance, the anhinga catches fish by running them through with its long, dagger-like beak. An anhinga is a water bird. Unlike most other water birds, it has neither air sacs to help it float nor oil glands to make its feathers waterproof. It lives along the warmer lakes and rivers of the southeastern United States.

Some waterfowl float so well they must struggle to feed under water. Not the anhinga. When it settles in the water, its feathers grow wet. It sinks instead of floats. Only its skinny neck and head stay above water. It can dart around underwater with ease. It is an excellent swimmer.

An anhinga can silently sink below the surface without a splash that frightens the fish that are its prey. Its black feathers make it nearly invisible underwater. As the anhinga swims, it pulls back its neck, ready to strike. A special hinged muscle allows it to thrust its head forward quickly. It spears a fish with its long, daggerlike beak. Then it brings the fish to the surface, tosses it into the air, and swallows it head first.

After a few minutes in the water, the anhinga's feathers become waterlogged. They no longer offer protection against the water. After diving for a meal, it comes to the surface and climbs onto a low limb by the water. Unable to fly with wet feathers, it spreads its wings to the sun. Its glossy black feathers absorb solar heat, drying them.

While waiting for the feathers to dry, the anhinga digests its most recent meal. After 20 minutes or so, the feathers are dry, and the bird is warm. It can dive into the water to spear fish again.

The flamingo has a beak for straining out small life. The Spanish explored the Caribbean islands in the 1500s. They saw many amazing sights, including large groups of wading birds with exceptionally long legs, a long, flexible neck, and an unusual down-turned beak. But most noticeable were the birds' bright reddish orange feathers. When the birds suddenly wheeled into the air, it appeared as if flames reached for the sky. They called the birds flamingos — flaming birds.

Flamingos thrive because of their design. With partially webbed feet, long legs, long neck, and special beak and tongue, they can eat food so small as to be almost invisible. In addition, they are content to eat from the bottom of marshes with water that is too salty for many other birds.

A flamingo makes a meal of tiny algae, snails, and shrimp. When feeding, it stirs up the water by stamping around in the mud with its feet. Then it puts its head and bill upside down below the surface. The bird gulps in the muddy water. Using its tongue, it forces the water and mud out. Tiny, comb-like structures along the bill trap the edible material.

Flamingos are a popular bird, and for some reason people like to put plastic pink flamingos in their yards — even in northern states that the birds seldom call home.

Most birds can digest food quickly. Their stomachs have three chambers. The first chamber is the crop. Parent birds store food for their young in the crop. After flying to the nest, they can bring the food back up from the crop to feed their young. Below the crop is the stomach that adds digestive juices to the food.

The third chamber is the gizzard. Birds have no teeth to chew their food and mix it with digestive juices. They use the gizzard to grind food. The gizzard has muscles surrounded by a tough lining. The bird picks up coarse sand and pebbles and swallows them. The gravel is kept in the gizzard. Muscles contract the gizzard to grind rocks against hard foods such as nuts and seeds.

Some birds eat seeds that are often coated with a hard shell and would pass through the bird's digestive tract if it were not for the gizzard. The grit in the gizzard mechanically grinds up seeds and makes them easier to be digested in the intestines.

The gizzard helps a bird achieve flight by reducing its weight. The gizzard is lighter than two large jaws and a mouth full of teeth. By using a beak designed for gathering food and a gizzard for crushing it, the bird is able to digest food quickly but does not have the added weight of heavy jawbones and teeth.

Many birds migrate. The record holder appears to be the Arctic tern. It breeds in the far north. Near the Arctic Circle during summer, the days are long and the nights are short. The weather is warm and food is abundant. But as the long night of winter approaches, the bird flies all the way south to the Antarctic, half a world away. Once again, it has nearly continuous sunlight. Each year, the Arctic tern flies from north to south and back again — a distance of 25,000 miles.

Why do birds migrate? Birds migrate for a variety of reasons. For instance, ducks need open water. As winter approaches, lakes and ponds begin to freeze all the way across. Ducks start the southward migration and stay ahead of the freezing weather. Other birds are not bothered by cold, but they do need a ready supply of food. They migrate to find food.

Another reason for migration is to reduce the number of predators. A predator such as a fox lives in the same area year round. A food supply must be readily available all year long. By flying away for part of the year, birds such as geese deprive foxes of food. With the birds gone and a harsh winter coming on, only a few foxes can find food and survive.

When birds return to the North, they find far fewer predators than if they had stayed in the area all year. Birds also flock together in vast numbers as a way to survive. The amount of food a fox can eat in a day is limited. Geese fly together and land in great numbers. Even the most successful predator is unable to eat all of the birds. Most birds in the flock will survive and reproduce to replenish the species.

Most Dangerous Bird

The cassowary (KAS-uh-WER-ee) is a large bird that lives in Australia and New Guinea. Standing more than six feet tall and weighing as much as 180 pounds, it is heavier and taller than an average human. Other large birds include the emu of Australia, the rhea of South America, and the ostrich of Africa.

The emu, rhea, and ostrich live in open country. They escape from enemies by running. Cassowaries live in rainforests, which makes running difficult. Cassowaries are equipped with a helmet-like, bony crest on their heads. The helmets allow them to crash through rainforest undergrowth without hurting their heads.

The male has a big role in raising young. He builds the nest, and after the female lays the eggs, the male takes over. He keeps the eggs warm and guards the nest. After the little ones hatch, the male spends nine months teaching them how to forage for food. Their main food is fruit that they pluck from branches or that has fallen to the ground.

Although cassowaries can run fast, should an enemy threaten them, they will turn and fight. A cassowary vigorously defends its young and can be dangerous when cornered. It protects itself by kicking with its powerful legs. The cassowary's three-toed feet have sharp claws. The middle claw is shaped like a dagger and is more than five inches long. It has caused injury and even death to humans. Many experts agree that it is the world's most dangerous bird.

Cassowary

Discovery

1. **Plumage of males, females, and juveniles differ and can also change during winter.**

2. **They both were killed for food.**

3. **Flying away leaves predators without a source of food, reducing their numbers.**

Questions

A B   1. Birds are (A. cold-blooded   B. warm-blooded).

A B   2. Compared to other vertebrates, birds have hearts that beat (A. less   B. more) quickly.

A B C D   3. The feature birds do NOT share with other animals is (A. a backbone   B. feathers   C. reproducion by laying eggs   D. walking on two legs).

            4. Birds grown for human consumption are referred to as _____.

            5. What birds did the British use to fly messages during the war with Napoleon? _____

A B C D   6. The person who made bird watching popular was (A. Carl Linnaeus   B. John J. Audubon   C. Roger Tory Peterson   D. The Duke of Wellington).

A B C D   7. The plumage of a species of a bird can vary depending on whether (A. it is a juvenile   b. it is male or female   C. it is summer or winter   D. all of the above).

A B   8. The brighter and more distinctive plumage usually belongs to the (A. female   B. male) bird.

            9. Why do birds eat seeds and insects rather than grasses? _____

A B C D   10. An anhinga can sink below the surface easily because (A. it is almost entirely without feathers   B. its feathers are black and heavy   C. its feathers are coated with oil   D. its feathers have no oil coating).

           11. Rather than teeth for crushing food, birds have a _____ filled with grit and small stones.

T F   12. One reason birds migrate is to find food.

13. Matching:
    __ Can mimic human speech
    __ Used to hunt small game
    __ Thick, heavyset, and flightless, this bird is extinct
    __ Once numerous in the United States, now extinct
    __ Has a long, thin beak that is like a hollow straw
    __ Has a long, dagger-like beak used to spear fish
    __ Migrates from Arctic to Antarctic
    __ Large bird capable of killing a human

a. anhinga
b. Arctic tern
c. cassowary
d. dodo
e. falcon
f. hummingbird
g. parrot
h. passenger pigeon

Explore More:

What are some common birds in your area? Can you name them? Watch them to learn about their daily life. Sketch two dissimilar birds and mark how they differ from one another — color, beaks, and type of claws.

The voice box of a bird is called a syrinx. How does it differ from the human voice box? Some birds can achieve a range of tones and have a distinctive call. What birds can you identify from their call or song alone? Some bird-watching books attempt to portray a bird's song as a musical diagram. Study such a diagram for birds whose call you know.

How does the structure of feathers help insulate birds? Why do birds molt?
Explore the flight characteristics of birds. Which ones use thermals to remain aloft? What are some typical speeds at which birds fly when they migrate? Why do geese fly in a V formation? How do they find their way when migrating? How do birds build nests? What do they use? How many eggs do different bird species lay?

Chapter 13

Mammals

Mammals are members of the subphylum Vertebrata that also includes fishes, amphibians, reptiles, and birds. All mammals are vertebrates — they have backbones. All maintain an internal temperature that does not depend on the environment — they are warm-blooded. Mammals have sweat glands to help rid their bodies of excess heat in warm weather. Most have hair or fur to keep them warm in cold weather. All mammals breathe air with lungs and have hearts with four chambers.

Mammals share many of their characteristics with other animals that are not mammals. But mammals do differ in one important way that sets them apart from all other animals: Mammals raise their young on milk.

The word *mammal* is from a Latin word referring to glands that produce

Explore

1. How do mammals differ from other vertebrates?
2. What mammals dig underground, swim underwater, and fly in the air?
3. What is the fastest mammal and largest mammal?

Mammals That Lay Eggs

Almost all mammals give live birth to their young. But two mammals lay eggs: the platypus and the spiny anteater.

The platypus is a mammal of Australia and the island of Tasmania. The female lays eggs in an underground den. She then curls around them to keep them warm. A platypus has webbed feet and a tail shaped like that of a beaver. The tail acts as a rudder to help it turn quickly when swimming. The male platypus has spurs on its back legs that are connected to poison glands. It is the only mammal that is venomous.

The spiny anteater is a mammal of Australia and the islands of Tasmania and New Guinea. It has a long, sticky tongue for catching and eating ants. The female spiny anteater lays one egg and places it in a fold of skin on her abdomen. After 10 days, the egg hatches. The mother carries the little spiny anteater in her pouch for nine weeks. When it becomes too large for her to carry, she builds a nest for it.

Both the platypus and spiny anteater produce milk for their young, which is why they are classified as mammals.

Mammal Classification

Kingdom: Animal
Phylum: Chordata
Subphylum: Vertebrata
Class: Mammals

milk. Lactose is a sugar found in milk. It is sometimes called milk sugar. Young mammals digest milk and convert lactose into glucose, also known as blood sugar. Glucose is the source of energy for cells. Human infants digest lactose easily, and convert it to glucose.

Milk is made of water, protein, sugar, and fat. It also contains minerals such as calcium and phosphorus. The amount of each one in milk varies depending on how the young use the milk. Seals and other mammals that dive in cold water have an insulating layer of fat. Mothers of young seals provide milk that is high in fat so the young can develop a layer of blubber. The milk of seals is about 40 percent fat. It is as thick as melted ice cream.

Some newborn mammals, such as horses and zebras, need to grow rapidly and run quickly. Newborn horses, known as foals, are on their feet within minutes after being born. Their mothers, mares, give milk that is only about 1 percent fat. In drinking the mare's milk, the foal receives a greater proportion of sugar for energy, calcium and phosphorus for building bones, and protein for growth.

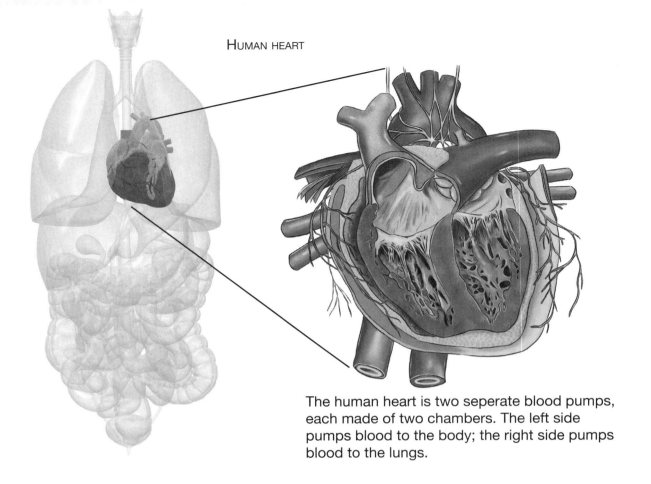

The human heart is two seperate blood pumps, each made of two chambers. The left side pumps blood to the body; the right side pumps blood to the lungs.

Besides giving milk, mammals have other traits. Some of the traits are shared with animals that are not mammals. All mammals are warm-blooded. However, they share that feature with birds. Female mammals and sometimes the males take active roles in protecting their offspring and training them. Both male and female birds also tend to their offspring until they are old enough to fend for themselves.

Mammals have four-chambered hearts, a feature they also share with birds and crocodiles, a type of reptile. A four-chambered heart is divided into two pumps that work at the same time. However, a wall separates the two sides of the heart. Blood does not directly mix between the two halves. The left portion takes blood from the lungs that is rich in oxygen and pumps it to the rest of the body. Cells remove the oxygen and dump waste products into the blood. Veins return the used blood to the right side of the heart. The right side pumps the blood to the lungs. There the waste gases are released and the lungs recharge the blood with oxygen. Except for mammals, birds, and crocodiles, other vertebrates do not have four-chambered hearts. They do not fully separate the oxygen-rich blood from blood that has gone throughout the body.

Sometimes a feature is true for the majority of mammals, but not for all of them. Most mammals are covered with hair or fur. Whales are an exception. Whales are mammals because they maintain a constant body temperature, have four-chambered hearts, give birth to live young, and nourish them with milk. Yet, whales are devoid of hair as adults.

The one trait that is always true for mammals is that they feed their newborns on mother's milk. Because the mother must provide the milk, there is a limit to the number of offspring. Mammals

Rodent Classification

Kingdom: Animal
Phylum: Chordata
Subphylum: Vertebrata
Class: Mammalia
Order: Rodent

seldom have more than 20 offspring at once. Many give birth to only one or two at a time. Because they have fewer offspring, offspring stay with the parents until they can fend for themselves. Many animals have a set of inborn rules for survival. These rules are known as instinct. But mammals can add to these rules by training their offspring with new information that the adults have learned.

Scientists have identified about 5,000 species of mammals. Other animals, such as insects, have far more species. For instance, scientists have classified dragonflies alone into about 5,000 species and grasshoppers into about 17,000 species. Despite the smaller number of species, mammals survive in practically every habitat. Polar bears thrive in arctic cold. Camels trek across dry and hot climates. Mountain goats climb the rocky summits of high mountains where the air is thin. Sperm whales dive to a depth of more than a mile in the ocean.

Mammals differ in size. Kitti's hog-nosed bat of South America is the smallest mammal. It is about the size of a bumblebee and weighs less than a dime. The largest mammal is the blue whale, with a length of 100 feet and a mass of 150 tons. It is not only the largest mammal, but also the largest animal that has ever lived, even bigger than the largest dinosaur.

Mammals differ in life span. The tiny shrew is common in the United States. A shrew looks like a small, nervous mouse, but with a long, pointed snout. Shrews live but 18 months, which scientists believe is the shortest life span of any mammal. Larger animals usually live longer. Elephants live about 70 years.

Mammals differ in where they search for food. Moles burrow under the ground, whales swim in the sea, and bats fly in the air.

Bats are the only mammals that can actually fly. Some mammals, such as the flying squirrel, climb high in a tree and launch themselves into the air. They glide to another tree at a lower level. Or, if they catch a strong updraft, they may be able to gain height. Bats can gain altitude in still air, and they are good fliers. In straight and level flight, bats can speed along at about 20 miles per hour. A bat's wing has a complicated arrangement of skin, bones, and muscles.

Bats are nocturnal (nok-TUR-nuhl), active at night or around the time of dusk or dawn. They sleep during the day. Bats spend the day hanging upside-down in caves.

The female bat usually gives birth to a single offspring. Because she must continue to hunt, the baby bat hangs diagonally across her abdomen to

Dog and Cat Classification

Kingdom: Animal
Phylum: Chordata
Subphylum: Vertebrata
Class: Mammalia
Order: Carnivore
Family: Canine (dog)
          Feline (cat)

keep her in balance while she flies. When the baby grows older and heavier, she leaves it behind, hanging upside-down in the cave while she goes hunting for food.

Some bats eat fruit, but others eat insects. The fruit-eating bats pollinate many tropical plants, including bananas, dates, and figs. They drink the juice of the fruit. A single bat may drink three times its body weight in juice in a single night.

Insect-eating bats feed on insects that they catch in the air. They identify the direction and location of the flying insect by echolocation. Bats emit a high-pitched squeal that cannot be heard by human ears. Their large ears detect the returning echo and sense the direction to the insect. They can fly quickly and turn sharply to catch the insect. When feeding, a bat can capture about two insects every second. In a single night, a colony of bats eats tons of insects.

Moles are designed for an underground life. Many mammals dig burrows and tunnels to escape predators and provide dens to raise their young. These mammals include prairie dogs, gophers, ground squirrels, mice, and marmots. Most spend only part of their time underground. They come to the surface for food.

However, some species of moles seldom come to the surface. Instead, they live underground. Moles eat the roots of plants in gardens. They have strong, bullet-shaped bodies and powerful, shovel-like front feet with claws for burrowing. Moles push through the soil rapidly, producing a ridge on the surface. They push out dirt to make vent holes for letting air into their tunnels.

Animals lose heat through their skin. A small animal has a greater proportion of skin compared to the volume of their bodies than a larger animal. They lose heat at a rapid rate. Because of their small size, moles are in constant need of food to maintain their internal temperature. Moles can only sleep for a few minutes before they become hungry. They must hunt day and night and take only short naps.

Many mammals live in, on, or near water. Seals, sea lions, and walruses do their hunting for food in the sea. They come to the shore to give birth. But whales, dolphins, and porpoises spend their entire life in the ocean. They give birth to their young in water.

Although whales live entirely in water, they must come to the surface for air. Whales breathe through a hole in the top of their heads called blowholes. Baby whales, called calves, are born underwater. But they must have air to breathe. The mother nudges the newborn calf to the surface for it to take its first breath.

Like all mammals, whales are warm-blooded. They maintain a body temperature around 99oF. Many whales feed in the cold water of the Arctic and Antarctic. The thick layer of blubber preserves the whale's body heat in cold ocean water.

A newborn calf would be unable to survive in such cold conditions. Instead, whales migrate to shallow, warm seas near the equator. There they give birth. Mother whales provide up to 160 gallons of milk a day to the calf. The young ones develop the layer of blubber that will let them hunt for food in colder water. When they go north to the feeding grounds, they gain even more fat. In addition to keeping them warm, the layer of fat helps them float effortlessly in water.

Blubber is a storehouse of energy. When it comes time to migrate, the whales convert the fat back to energy for the long swim back to warm climates.

The mammals we are most familiar with are those that live on the land. And of the land animals, rodents make a sizable portion. Almost half of all mammal species are rodents. Rodents include mice, rats, squirrels, chipmunks, prairie dogs, beavers, hamsters, and guinea pigs. What they have in common is their teeth. All rodents have large, sharp front teeth designed for gnawing. But they do not have canine teeth.

With their prominent front teeth, rabbits would appear to belong to the order rodent, too. But that is not the case. Rabbits have six front teeth, four on the top and two on the bottom. At first glance, they appear to have two on top and two on bottom. Looking more closely reveals that they have two teeth hidden right behind the top ones. These teeth, called peg teeth, are not found in rodents.

The word *rodent* is from a Latin word meaning "to gnaw." Rodents use their two front teeth, incisors, for cutting. Beavers cut down trees with their sharp teeth. Squirrels and rats can cut through the insulated wiring found in homes.

Rodents' teeth maintain a sharp edge because of the way they are made. In front is hard enamel. In the back of the tooth is softer dentine. As the rodent gnaws, the back dentine wears away so only the sharp enamel is left. The tooth actually gets sharper with use.

The front teeth of rodents, both upper and lower, constantly grow. Some rodents live on soft food. Rodents gnaw even when not looking for food to keep their teeth sharp and wear them down. If a beaver did not wear down his constantly growing teeth, his teeth would be four feet long

RODENTS

Mouse   Chipmunk   Hamster   Squirrel   Beaver

after a year.

Rats and mice are rodents that are considered pests. They eat grains in warehouses. For people of long ago who struggled to survive on their crops, the loss of grain to rats and mice was a real problem. One way they controlled the rats and mice was with animals such as dogs and cats. The dog known as a rat terrier was specifically bred for controlling vermin such as rats.

Mammals such as the rat terrier that kill and eat other animals are in the order Carnivore. Carnivorous animals include the dog, cat, wolf, and bear. They have several characteristics that help them catch prey. They have well-developed senses. They have sharp eyes, sensitive ears, and a well-developed sense of smell. They can pounce quickly and run fast. Carnivores are also intelligent. Sometimes they work with one another to bring down a larger animal. Wolves, for instance, hunt in packs.

Carnivores have special teeth for holding their prey. These teeth, called canines, are larger than the others and are located on both sides of the front teeth. The canine teeth are long and pointed. They project down below the other teeth. They help the animal grasp its prey. The teeth are named after the Latin word for dogs, in which the teeth are prominent.

Cats have canine teeth, too, and are in the

Big Cat

The tiger is the largest member of the cat family. It lives in Siberia, India, and Southeast Asia. The large and rare Siberian tiger can weigh 640 pounds and be 10 feet long, not including the tail. Unlike most other cats, tigers swim well, and they appear to enjoy being in water.

The roar of a tiger is especially loud and can be heard two miles away. The loud roar can cause a hidden animal to reveal itself. The frightened animal bolts from its hiding place. The sound vibrations seem to come from all directions, so sometimes the confused animal will run toward the roar rather than away from it.

A tiger has awesome power and will attack practically any animal except elephants and rhinoceroses. A good meal for a tiger is about 75 pounds of meat.

The best-known poem about the animal is *The Tyger*, written by William Blake in 1794:

> Tyger, tyger, burning bright,
> In the forest of the night,
> What immortal hand or eye
> Could frame thy fearful symmetry?

He used *y* instead of *i* to spell the name. Not until the 1850s was "tiger" spelled as it is today. The word *symmetry* in the poem refers to the fact that the stripes on the left side of the tiger's body are exactly like those on the right side. The stripe patterns are unique to individual tigers. Like fingerprints in humans, stripes can be used to tell individual tigers apart.

Sheep

Elephants are big, cheetahs are fast, and tigers are ferocious. Sheep, on the other hand, are meek animals. Yet sheep are more numerous than the bigger and stronger animals.

People have raised sheep since ancient times. The first farm animals named in the Bible are flocks that Abel kept (Gen. 4:2). Abraham was described as wealthy because he owned sheep and cattle (Gen. 24:35). In the New Testament, shepherds who were tending their flock at night learned about the birth of Jesus (Luke 2:8–12). Even today, in some places sheep outnumber the human population. Australia has eight times as many sheep as people. In North America, the state of Wyoming has more sheep than humans.

Despite their numbers, sheep are not particularly strong. They are easily frightened and feel safer in a flock. They depend upon shepherds to take them to pasture and protect them from enemies such as wolves, coyotes, and mountain lions.

Sheep, like elephants, eat grass. Sheep have no top front teeth. This design allows them to nibble grass very close to the ground. It prevents them from pulling up plant roots and destroying the lands they graze.

A female sheep is a ewe, pronounced "you." The females give birth to lambs in the spring. They usually have one or two lambs. If more are born, then a ewe without a lamb will adopt the extra lamb. Sheep have poor eyesight but a good sense of smell. Mothers and their lambs recognize one another by the smell and sound. Sheep make a *baa-baa* bleat. A baby lamb can identify its mother by her bleat.

As cold weather approaches, sheep grow a warm winter coat of wool. After winter, farmers shave off the thick fleece of adult sheep. Sheep bred for their fine wool account for nearly half the world's sheep population. The wool is very fine and soft.

Once shorn from the sheep, the bags of wool are sent to a factory to be made into sweaters, suits, and other items of clothing. Wool is made of a type of protein similar to the one that makes your fingernails. For that reason, it is flame resistant. Wool also provides excellent protection from cold and wet weather. It holds in body heat even when wet.

Here is a nursery rhyme that tells about wool:

Baa, baa, black sheep,
Have you any wool?
Yes sir, yes sir,
Three bags full.

One for the master,
One for the dame,
And one for the little boy
Who lives down the lane.

same order as dogs. Although cats and dogs are in the same order (Carnivore), they are in different families. The dog belongs to the canine family that includes animals such as wolves, foxes, and jackals. The cat belongs to the feline family that includes such animals as cheetahs, lions, tigers, and jaguars.

Cats have soft fur, padded feet, claws that

retract when not in use, a rough tongue, and sharp teeth used for grasping and tearing. They have good hearing and especially good night vision. All cats can purr, and the larger ones can roar. They are quick on their feet.

All cats are speedy runners, but the cheetah is faster than any cat or any other land animal. People in India gave the cheetah its name. They called it *chita*, a word that means "spotted one" in their language. A cheetah's sandy yellow coat is covered with small, round black spots. Its face has tiny spots that run from beneath its eyes and down its cheeks to the corners of its mouth. The spots look like teardrops, so a cheetah always appears sad, as if it has been crying.

The cheetah lives in the dry grasslands of Africa and southwest Asia, where it hunts grazing animals such as antelopes. It hunts during the day. It carefully creeps forward toward its prey. Then when it gets within 100 feet, it gives a burst of speed to catch the antelope.

A cheetah is the fastest animal on land. It can reach a speed of 60 miles an hour, about 88 feet a second. The cheetah makes about four strides a second, and each stride covers 22 feet. Measure 22 feet, and you will see that it is indeed a long distance. It gets up to speed quickly, too, in only two or three steps.

Cheetahs are cats, but they differ from other cats in some ways. Although they can purr, they cannot roar as the lion or tiger. They communicate to their young or other cheetahs with barklike sounds and birdlike chirps.

Most cats hunt at night, but cheetahs hunt during daylight hours. Many cats pounce on their prey with a sudden leap. Rather than jumping ability, cheetahs depend on their speed. A cheetah's running muscles need plenty of oxygen, so it has enlarged air passageways, a large heart, and big lungs. Its liver stores blood rich in oxygen to release during the chase.

A cheetah has a flexible backbone. During a high-speed chase, the belly muscles tighten and make the spine arch like a bow. The spring of the spine is released into the next step. The cheetah runs with its belly muscles as well as with its leg muscles. A long, flexible tail acts like a rudder to improve balance for high-speed turns.

When running, its stride is so long that its back legs extend in front of the cheetah's head. Its front legs would be in the way, but the cheetah crosses its front legs around its chest so they are out of the way of the back legs. Then their front legs drop down and extend the stride as the back

legs pass. All of this happens so quickly that it is only clearly visible in slow motion movies of the animal as it runs.

The cheetah cannot keep up its fast pace for very long. If the animal it chases manages to elude it for a quarter of a mile, the cheetah is done. The cheetah has no sweat glands to cool itself. During long runs, its body becomes dangerously overheated.

But the cheetah chooses it target wisely. More than half of the time, it catches a meal on its first try. Usually, the cheetah falls exhausted by its catch. It is too weak to eat until it has rested and cooled down. While it is so exhausted, hyenas or lions may try to steal its meal. Two or three cheetahs from the same litter may hunt together and protect the food from thievery.

The elephant is the largest land animal. At one time, elephants were in an order called *Pachyderm* (paki-durm), along with rhinos and hippos. Because of changes in the classification, pachyderm is no longer a scientific classification. Instead, the elephant is a member of the Proboscidea (PRO-boss-ID-ee-uh), an order that is made of elephant, mammoth, and mastodon. The mammoth and mastodon are extinct, so the elephant is the only surviving member of that order. Proboscidea is from a word that means "long, flexible nose."

Despite the changes that scientists make, people still refer to elephants as pachyderms. The name comes from Greek words *pakhus*, meaning "thick," and *derma*, meaning "skin." A pachyderm is an animal with a thick skin. The name certainly applies to an elephant. An elephant's skin is about 1.5 inches thick. The skin is sensitive, too, because elephants can feel insects that craw around on them.

The African elephant is the largest land animal. A large one weighs more than 14,000 pounds — about the same as four full-size automobiles. Elephants continue to grow throughout their lives, so the longer they live, the larger they get.

Such big animals need a lot of food. Each elephant eats about 500 pounds of food a day. They eat grass, roots, bark, twigs, and fruit. If a herd of elephants were to stay in one place, they would quickly consume all the food in the area and kill the plants on which they grazed. Instead, they move from day to day and travel 3,000 miles or more in a year. Because they keep moving, they never destroy the food

Elephant Classification

Kingdom: Animal
Phylum: Chordata
Subphylum: Vertebrata
Class: Mammalia
Order: Proboscidea

supply in one area, and by the time they return, it has recovered enough for them to have another meal.

The leader of a herd of elephants is usually the oldest female, called the matriarch, which means "mother." Young adult males act as scouts. When the herd comes to an open area, scouts go out first and explore for danger before the rest of the herd follows.

Humans have used trained elephants to haul heavy loads since ancient times. No one knows for sure how the elephant got its name. At first, elephant was spelled several different ways in the English language. A book written in 1398 states, "The elyphaunt hath a longe nose lyke a trumpe." (The elephant has a long nose like a trumpet.) Not until the 1600s was the word spelled the way we write it today.

The elephant's trunk is amazing in its flexibility and usefulness. It is powerful enough to lift a tree but delicate enough to pick up a small object the size of a coin. The elephant uses its trunk for pulling down leaves to eat, for squirting water into its mouth, and for breathing when submerged under water.

The trunk has about 150,000 muscles.

Learning how to use it takes time and training. A baby elephant, called a calf, needs to practice for six months before it learns how to squirt water into its mouth with its trunk.

Discovery

1. Mammals provide milk for their young.

2. Moles live underground, whales inhabit the oceans, and bats fly in the air.

3. The cheetah is the fastest mammal, and the elephant is the largest mammal on land.

Questions

A B C D 1. The one that is NOT true of mammals is that (A. mammals are cold-blooded   B. mammals are vertebrates   C. mammals have a four-chambered heart   D. mammals have sweat glands).

T F   2. The one feature that is true of all mammals is that the females provide milk for the young.

A B   3. The milk that contains more fat is the milk of (A. horses   B. seals).

T F   4. A four-chambered heart is actually two separate blood pumps.

5. Why are the platypus and spiny anteater, who lay eggs rather than give live birth, classified as mammals? _____

6. The mammal that can fly is the _____.

A B   7. The word *nocturnal* means (A. active at night   B. hunt by sound echo).

8. An example of a mammal that spends most of its time underground is the _____.

T F   9. Whales must come to the surface to breathe air.

A B   10. Animals such as squirrels and beavers are examples of (A. canines   B. rodents).

A B   11. Cats are mammals, in the order carnivore, and belong to the family (A. canine.   B. feline).

A B   12. Cheetahs catch their prey primarily by (A. pouncing on them with a sudden leap   B. chasing them down).

A B   13. Elephants are still often referred to as pachyderms because of the physical characteristic of (A. thick skin, B. long, flexible nose).

T F   14. Both sheep and elephants eat grass.

Explore More

All animals except mammals and birds are cold-blooded. What advantage does a warm-blooded design have over a cold-blooded design?

Bears are mammals that hibernate during winter. What is hibernation, and what advantage does it give bears?

Some predators will only eat animals they have killed themselves. But some mammals, such as hyenas, are not as fussy. They will eat carrion, the remains of animals that have died or been killed by other animals. What role do these animals have in nature?

In addition to the platypus and spiny anteater, Australia has a large number of marsupials, such as kangaroos, wombats, koalas, Tasmanian devils, and Tasmanian wolves (extinct). The opossum is the only marsupial that lives in the United States. Investigate these animals to discover how marsupials differ from other mammals.

Chapter 14

Frauds, Hoaxes, and Wishful Thinking

The word *science* means "to know." Like all scientists, biologists strive to reveal the truth about the subject of their study. They often come back to the same subject time and again to learn more and correct past mistakes. They write reports about what they have learned. Biologists read the reports and then check one another's work to ensure that it is as accurate as possible.

Biologists assume errors are not deliberate. They shudder at the thought that a biologist might willfully report evidence as true knowing the information is false. But people do intentionally mislead others. They do it for money, fame, or merely in an attempt to impress others or embarrass them.

Explore

1. What is a fraud?

2. What is a hoax?

3. What amazing discovery did a 12-year-old girl make?

A fraud is a deliberate deception designed to gain money or something else of value. As an example of a fraud, and the damage it can do, consider the fossil known as the feathered dinosaur.

Paleontologists (PAY-lee-on-TOL-uh-gistz) are biologists who study forms of life of the past. They study fossils. In October 1999 paleontologists called a news conference to reveal a new fossil. They posted photographs of the fossil and showed realistic models of the animal as they imagined it must have appeared. The ancient animal had a birdlike bone structure and a strong, dinosaur-like tail.

The animal based on the fossil was given a scientific-sounding name — *Archaeoraptor liaoningensis*. The word *arch* means "ancient" and *raptor* means "bird of prey." Liaoning was the province in China where the fossil had been found.

Some paleontologists claimed that it was a missing link between dinosaurs and birds. The idea of a feathered dinosaur had been proposed before, but the majority of bird experts dismissed the idea. But this new fossil appeared to show otherwise.

Scientists began to investigate the fossil. It had come from China and made its way to the United States. China had laws against removing fossils from the country. Fossil hunters had somehow managed to smuggle it into the United States. It first appeared in a rock and gem show in Arizona, where it sold for $80,000. Next, it was donated to a small dinosaur museum in Utah.

Scientific magazines hesitated to print articles about the fossil, because it had come into the United States illegally. But the museum director agreed to return it to China, although he hoped to keep it on display for a few years. Chinese officials sent a fossil expert named Xu Xing to arrange for

Archaeoraptor liaoningensis

its return. Xu Xing worked at a museum in Beijing, China's chief city.

The fossil had been revealed by splitting open the original layer of rock. The split would have produced two pieces of stone, each bearing an imprint of the fossil that was between them. Each side of the rock would have identical impressions of the fossil. The American fossil museum had one side of the stone. Xu Xing looked for the other slab. He traveled to Liaoning Province, where the fossil had been uncovered.

As he did this, experts in the United States carefully studied the photographs of the fossil.

From the photographs alone, they raised serious questions. The bones between the main body and the tail did not match.

Within two months, Xu Xing had found the other side of the tail slab in a private collection in China. What he saw left him baffled and then deeply concerned. The China section did not match the one in the United States. The two should have been identical, but the one in China had a different shape. The more he studied the fossil, the more suspicious he became.

Xu Xing sent an email to the American paleontologists. "I have very bad news," he wrote. "Though I do not want to believe it, archaeoraptor appears to be composed of a dinosaur tail and a bird body."

The truth came out. A farmer earned extra income as a fossil hunter. He had found two fossils and glued the slab of the dinosaur tail to the lower portion of the birdlike body. He used bits and pieces of other fossils to fill in the gaps. The whole thing had been put together with homemade paste. He made the changes so his fossil would be more valuable.

The news caused scientists to examine other fossils. They were dismayed to see that other specimens had been altered to make them more attractive to buyers. The sellers did it for money.

Although the truth had come out, the damage had been done. The public had been deceived. In the case of the feathered dinosaur, some museums had put models of the mythical animal on public display. Television programs and network news featured stories about the feathered dinosaur.

Archaeologists study past human life and how people lived by examining old buildings, tools, pottery, bones, or other remains.

The ancient ruin of the Eurialo Greek Castle

Prehistoric pot

Bronze Luristan arrowhead, 500–1,000 BC

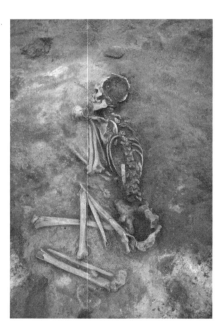

Ancient grave

Other museums had created displays that were intended to show a close connection between dinosaurs and birds. One museum displayed a miniature model of a dinosaur skeleton along with the actual skeleton of a large bird. The display stated that the two were very similar. Actually, they were vastly different. For instance, the bird had a strong breastbone that anchored its powerful flight muscles that were connected to its wings by tendons. The dinosaur had a weak breastbone. The dinosaur had a long, flexible tail, while the bird had a short, solid bone that held the tail feathers. This display and others were not removed when the original fossil was proven a fake.

Because it had been cobbled together with the goal of making money, the feathered dinosaur fossil was an example of a fraud. Biologists must guard against false ideas that capture their imaginations and override facts. When they fail to do so, they too readily accept frauds, fakes, and hoaxes.

A hoax is an elaborate prank intended to deceive but without the intent to gain money. The Piltdown man succeeded in fooling archaeologists (ar-kee-OL-uh-justz) for 40 years. Archaeologists study past human life and how they lived by examining old buildings, tools, pottery, bones, or other remains. Most scientists believe the Piltdown man was a scientific hoax. Apparently, no one made money from the deception.

Charles Dawson was a country lawyer and an amateur fossil hunter. In 1912 Dawson told his scientific friends an interesting story. He explained that he had been walking along a country lane in Piltdown Commons, Sussex, England. He stopped to watch a road crew working in a gravel pit. A workman threw out a reddish object. Dawson picked it up. "What's this?" he asked a workman.

"A coconut, sir," the workman said.

Dawson disagreed. "Stop! It's part of a skull."

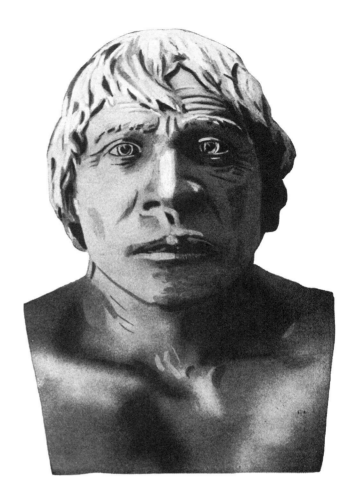

He claimed to have examined the gravel pit and found bone fragments from the top of a skull. The lawyer took the bones home to examine them more closely. Iron oxide in the soil had stained them dark red.

Dawson took the skull pieces to Dr. Arthur Smith Woodward, head of the Department of Geology at the British Museum. The skull immediately interested Woodward. It appeared very old.

Woodward and Dawson returned to the gravel pit and searched for other parts of the skeleton. They came across a peculiar-looking jawbone. It had teeth that appeared to be human, but the jawbone was shaped like an ape's jaw.

Woodward became excited about the discovery. He failed to notice that the jawbone had been stained by potassium dichromate, a chemical not normally found in soil in that part of England. It seems odd today, but no scientist questioned how the bones had been stained with the chemical.

The strange shape of the jaw convinced Woodward that the skeleton was half-man and half-ape. Another English scientist, Sir Arthur Keith, an expert on the brain, said, "The skull represents an extremely ancient form of man. It is the most primitive human brain so far uncovered."

Fifty years earlier, Darwin had proposed his theory of evolution. Some scientists looked for a direct link between human beings and subhuman forms. None had been found. Up until that time, all ancient bones had either definitely belonged to an ape, or definitely belonged to a man. This was the only skeleton that appeared to be a mixture of both.

Woodward gave the Piltdown man the scientific name *Eoanthropus dawsoni*, which means "Dawson's dawn man." Some prominent English scientists eagerly endorsed Woodward's claims about the discovery.

The new discovery burst upon the scientific world. Scientific journals and textbooks published models of the Piltdown man. The 1926 edition of the Encyclopaedia Britannica carried an entire article about the skull. Museums throughout the world put plaster casts of the Piltdown skull on display. The British Museum proudly displayed the actual Piltdown skull with the caption "Dawn Man."

Yet the whole thing was a hoax. Piltdown man never existed. In 1953 Kenneth P. Oakley and a team of researchers at the British Museum discovered the truth. Other scientists had worked with plaster casts of the bones. Oakley tested the actual bones. First, he used fluorine to find out the age. Fluorine slowly soaks into bones. An older bone contains more fluorine than a more recent one.

Fluorine dating showed that the top part of the skull had come from a skeleton of a person that had probably died during the Middle Ages, perhaps about 800 years ago. The jaw was that of an orangutan and only a few years old. Piltdown man was not ancient after all.

Next, he looked at the color of the bones. All were dark brown in color due to iron oxide rust. The color was only surface deep. Truly old bones

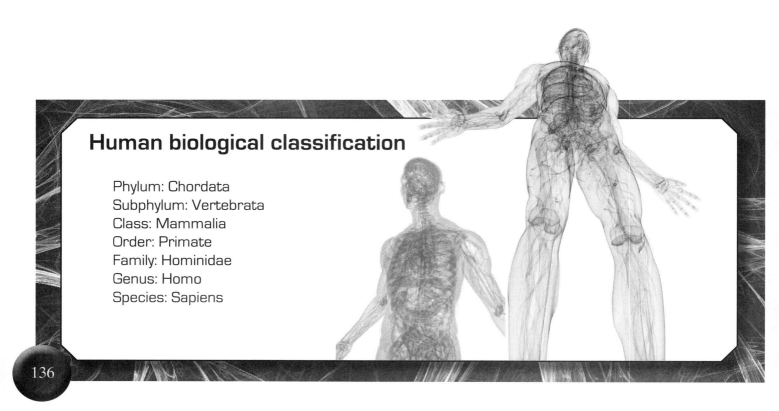

Human biological classification

Phylum: Chordata
Subphylum: Vertebrata
Class: Mammalia
Order: Primate
Family: Hominidae
Genus: Homo
Species: Sapiens

would have been stained through and through. The bones still had collagen easily visible with a microscope. Collagen, a fiberlike connective tissue, decays from bones as they age.

The teeth of Piltdown man were black, but Oakley scraped away the color. The black substance was paint. Underneath, the teeth still gleamed white. The teeth showed heavy signs of wear — like human teeth. Monkeys, apes, and orangutans eat tender shoots of grass and fruit that do not wear down their teeth. Humans eat a variety of food and their teeth wear down. Someone had filed the teeth to make them look human.

Kenneth P. Oakley's discovery touched off a wild uproar. The British Parliament demanded an investigation. Scientists who had defended Piltdown man tried to hide their embarrassment. They refused to speak to reporters. The painful hoax caused a terrible scandal.

Scientists questioned who produced the hoax and why. For years, the lead suspect had been Charles Dawson. However, those who knew him said he was an honest man. He could not be questioned because he had died in 1916, long before Oakley revealed the hoax.

In 1996 officials found a trunk stored at the British Museum. The trunk belonged to Martin Hinton, a volunteer who had worked at the museum 84 years earlier. When opened, the trunk contained bones that had been treated like those found in the gravel pit. An investigation revealed that Hinton had appealed to Dr. Arthur Smith Woodward for a full-time job at the museum. Woodward had turned him down. Some people suggested that Hinton may have been angry with Woodward and hidden the bones at Piltdown. He may have hoped to disgrace Woodward.

Regardless of the identity of the culprit, the hoax embarrassed not only Woodward but also the entire scientific community.

Another way that biologists can be led astray is by wishful thinking. Sometimes they allow their own views to sway how they interpret facts. Biologists can embrace incorrect ideas and be slow to reject them.

For instance, in 1856 workers in the Neander Valley near Duesseldorf, Germany, uncovered an ancient skull and bones. Rudolf Virchow, a well-known scientist, examined the bones. He was a medical doctor who had examined many dead bodies. His years of experience made him especially well qualified to examine the bones.

Virchow applied his skill to the ancient skeleton. He concluded that the bones belonged to a human being. The individual appeared to be short and powerfully built. Human beings do differ considerably in their body type. Pygmy people in central Africa along the equator are short. Most are no taller than five feet. Masai people, who live in the dry, sunny lands of eastern Africa, have tall, lean bodies. Masai can be more than seven feet tall. The bones Virchow saw fell within the range of human variation. In 1859 he stated that the bones were from a short but stout human being.

The next year, Darwin's theory of evolution took the scientific world by storm. Some biologists began a determined effort to prove Darwin correct. To prove Darwin's hypothesis correct, an ancient man would need to appear more savage than human beings of today. These biologists who embraced Darwin's idea competed with one another to find a missing link between humans and lower forms of life.

Skeletons similar to the one Virchow

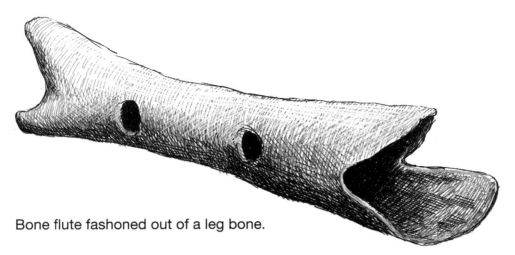

Bone flute fashoned out of a leg bone.

examined became known as Neanderthal (nee-AN-der-thawl) Man. French scientist Marcellin Boule made a model of the Neanderthal Man. He showed the legs bowed, back bent, and shoulders drooped. Boule's model was short and awkward.

Newspapers, textbooks, and cartoons portrayed Neanderthals as short, stoop-shouldered savages with thick hair and dim-witted minds. They were often called cavemen. The man had large, clumsy-looking hands.

Over the years, new discoveries cast doubt on the savage and apelike nature of the Neanderthals. Rather than cavemen clubs, scientists found a whole range of tools. There were spear points, stone lances, and personal decorations, such as pierced pendants.

Scientists also found a bone flute by one skeleton. The man had fashioned it out of a leg bone of a bear. He drilled four finger holes for making a variety of notes. At first, scientists thought Neanderthals had clumsy fingers entirely incapable of playing a musical instrument. But a computer model of the finger bones showed that a Neanderthal hand had the full range of motion. The Neanderthal could have played a tune without any problem.

Although more skeletons were found, none were put together to make a complete individual for almost 100 years. The first model by Boule remained the way Neanderthals were pictured. Finally, in 1957 a skeleton was put together properly. The people did not have an apelike appearance. Instead, they stood upright. They walked with a normal gait.

About 50 years later, researchers in Portugal found the ancient skeleton of a child. The boy had lived thousands of years ago and had been about four years old when he died. The bones had features showing that one of his parents was Neanderthal and the other a human. Neanderthals were of the same species as humans. Explorers in Romania confirmed this in 2002 when they found bones from three more skeletons with blended features.

Scientists today classify Neanderthals as *Homo sapiens*, the same as modern humans. *Homo sapiens* means "wise man." The Neanderthal story shows how incorrect ideas can stay firmly entrenched in science for a long time. Scientists must always test new ideas and be willing to change.

Usually, scientists are quick to accept information that fits in with already accepted scientific beliefs. Sometimes new information comes to light that appears to contradict older, more firmly established ideas. Should this happen, scientists are very reluctant to accept the new discovery.

For instance, many people, including some archaeologists, believe that ancient people were somehow inferior to modern man. Yes, ancient

Part of cave paintings seen in Matobo National Park near Bulawayo, Zimbabwe

people had to struggle for their food, clothing, and shelter. For most cultures, ordinary individuals had less time to devote to pursuits such as art. But when they did, they were every bit as talented as people of today. This fact is illustrated by the paintings in a cave in Spain.

In the northern part of Spain, a hunter discovered a cave in 1868. The cave, called Altamira, was on the estate of Don Marcelino de Sautuola. In 1875 Don Marcelino explored the cave. For four years he worked off and on in the cave. He found bits of bones, shells, and ashes from an ancient fire.

During the summer, Don Marcelino's 12-year-old daughter Maria asked to go with him into the cave. She entered a side cave. Suddenly she called to her father, "Papa! Come look! Painted bulls!"

Spectacular drawings of painted animals appeared to thunder across the ceiling. Don Marcelino had been in the chamber many times before. He had to stoop to go inside. Then he dug around in the dirt floor. He had never thought to look overhead.

The paintings amazed him. Animals appeared to move as if they were alive. Bison grazed, wild horses pranced, and angry bulls charged. The artist had used the bulges and hollows in the rocky roof to form rippling muscles.

Don Marcelino became fascinated by the cave art. He had accurate drawings made of the paintings. He published a book describing the paintings in 1880. The experts who

Ancient people painting cave by firelight

read the book could not accept his discovery. "A hoax!" they cried. "Primitive man was a barbarian and hardly more than an ape."

Don Marcelino's cave paintings showed that ancient people had great artistic ability. But to admit this would be to admit progress in reverse. Those who believed in the steady upward climb of mankind had their beliefs shaken by Don Marcelino's book.

In 1883 Don Marcelino appeared before a conference on prehistory. He showed the professors lumps of pigment. He passed around clamshells the ancient artist had used to hold the paint. He demonstrated how the artist had ground iron oxide and mixed it with animal fat to make oil paint in colors varying from yellow to dark orange and brown. Black pigment was made from charred bones.

Édouard Cartailhac, the leading professor of prehistory in France, denounced the paintings as frauds. The conference decided that the paintings were the work of a modern painter. Some scientists accused Don Marcelino himself of masterminding a hoax. They accused him of hiring an artist to secretly paint the cave ceiling. Rather than ancient art, the paintings were recent clever fakes, they claimed.

Three years later another conference met. The professors refused to let Don Marcelino — an amateur — appear before them. No scientific publication accepted his paper about the Altamira cave paintings.

The matter stood suspended in disbelief for 12 years. Then, in France, scientists opened a second cave. The entrance had been sealed for centuries. Inside, explorers found more cave art. The animals looked as alive as any painting in a Paris museum.

This time even Professor Cartailhac admitted he had been wrong. Ancient humans had created the vivid animals charging across the ceiling. He traveled to Spain to see the paintings and to apologize. Don Marcelino had long since died. The professor apologized to Maria instead. "It is I, and not your father, who have been foolish."

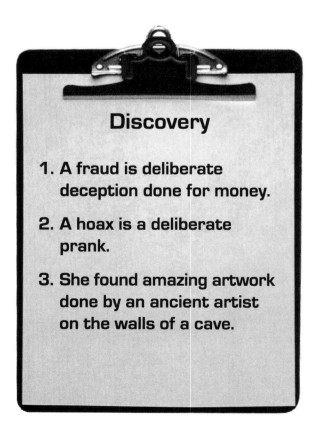

Discovery

1. A fraud is deliberate deception done for money.
2. A hoax is a deliberate prank.
3. She found amazing artwork done by an ancient artist on the walls of a cave.

Questions

A B   1. A deliberate deception designed to gain money or something of value is a (A. fraud   B. hoax).

A B   2. A paleontologist studies (A. past life   B. how animals migrate).

A B C D   3. The man who glued together parts to make a feathered dinosaur fossil did it: (A. to embarrass his employer   B. to gain entry into the United States from China.   C. to make the fossil more valuable   D. to prove his skill as a fossil hunter).

A B   4. When Xu Xing, the Chinese scientist, found the other fossil slab in China, he discovered that it (A. did not match   B. was in far better shape) than the one in the United States.

T F   5. The fact that the feathered dinosaur fossil was a fraud was discovered before it was made public.

T F   6. Archaeologists study past human life.

A B C D   7. The Piltdown man deception began with the discovery of: (A. broken pottery found in a trunk   B. detailed cave paintings   C. part of a skull   D. the remains of a leg-bone).

A B C D   8. The phony bones of Piltdown man were uncovered in: (A. England   B. France   C. Germany   D. Spain).

T F   9. Robert Virchow described the bones found in the Neander Valley as being from a short but stout human being.

A B C D   10. The classification of humans, *Homo sapiens*, means: (A. caveman   B. dawn man.   C. upright man   D. wise man).

T F   11. The only tools ever found by the Neanderthals were stone clubs.

12. Why did Maria see the painted bulls on the ceiling, but her father had overlooked them?

EXPLORE MORE — HOAX, FRAUD, FAKE, OR WHAT?

A visitor walked into a small trading post in the desert southwest of the United States. A display showed small pieces of sandstone with colorful designs on them. A sign over the small stones read, "Authentic Native American petroglyphs." A petroglyph (PET-ruh-GLIF) is a carving or drawing on rock. Petroglyphs made by prehistoric Native Americans are often found on government land. Removing them is illegal. Those from private property are rare and often expensive.

If someone had made fake Indian petroglyphs to sell, would the rock drawings be an example of a fraud or of a hoax?

The petroglyphs on display had a price of only one dollar. The visitor decided to purchase one. When he examined the back of his purchase, he noticed that the artist had signed his name. He looked at others and found autographs on them. In addition, each piece had a year and date written below the name. The date was less than a week old.

The visitor asked the shopkeeper if the small drawings were truly authentic Native American petroglyphs. The shopkeeper said yes, but he smiled and told the visitor to walk outside to the back of the building. There, in the shade, the visitor saw an art teacher and several Native American children working at a table. They drew the rock paintings.

Question: Do you think the rock drawings qualified as authentic Native American petroglyphs?

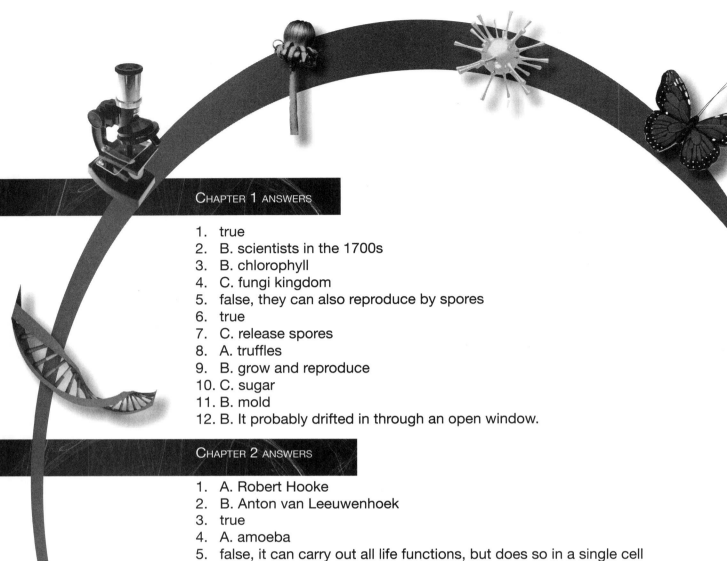

Chapter 1 answers

1. true
2. B. scientists in the 1700s
3. B. chlorophyll
4. C. fungi kingdom
5. false, they can also reproduce by spores
6. true
7. C. release spores
8. A. truffles
9. B. grow and reproduce
10. C. sugar
11. B. mold
12. B. It probably drifted in through an open window.

Chapter 2 answers

1. A. Robert Hooke
2. B. Anton van Leeuwenhoek
3. true
4. A. amoeba
5. false, it can carry out all life functions, but does so in a single cell
6. malaria
7. B. had chlorophyll
8. B. diatoms
9. true
10. fungi
11. false, anaerobic bacteria can survive without oxygen
12. B. nitrogen compounds in the soil
13. f. virus
    a. animal
    c. fungi
    d. plant
    e. protista
    b. bacteria

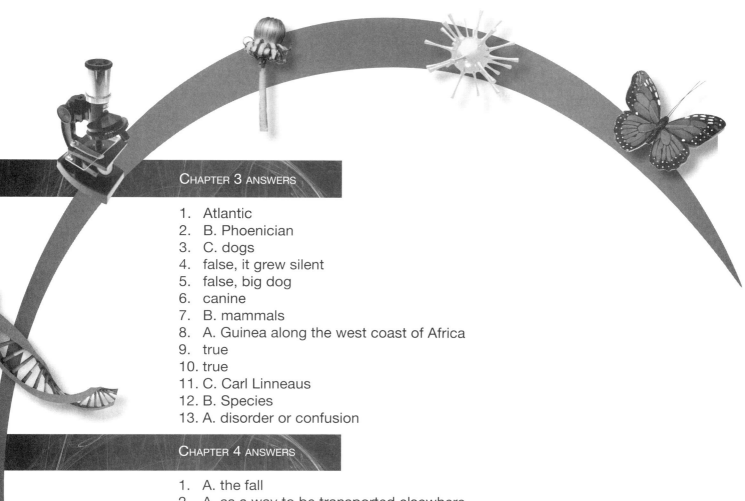

Chapter 3 answers

1. Atlantic
2. B. Phoenician
3. C. dogs
4. false, it grew silent
5. false, big dog
6. canine
7. B. mammals
8. A. Guinea along the west coast of Africa
9. true
10. true
11. C. Carl Linneaus
12. B. Species
13. A. disorder or confusion

Chapter 4 answers

1. A. the fall
2. A. as a way to be transported elsewhere
3. almonds
4. A. coconut
5. true
6. vegetative
7. B. do not flower
8. true
9. biennials
10. stem
11. false, sugar (glucose)
12. B. coffee

Chapter 5 answers

1. B. cereal grains.
2. False, in use by Native Americans before then
3. A. oxidation
4. A. fats
5. A. carbohydrate
6. bacteria
7. C. lactose
8. glucose
9. carbon, hydrogen, oxygen
10. true
11. B. heat energy
12. B. fat
13. D. proteins
14. B. proteins

Chapter 6 answers

1. B. mechanical
2. B. grazing animals
3. C. mash food
4. A. sugar
5. true
6. B. chemically changing the food
7. false, milk digests more quickly than a pickle, for instance
8. A. hydrochloric acid
9. B. small intestine
10. C. small intestine
11. A. amino acids
12. A. bitter taste

Chapter 7 answers

1. A. used a compost hotbed
2. B. planting a piece of a potato with an eye in it
3. true
4. California
5. plum
6. A. almond tree
7. Appleseed
8. He was too small and frail for heavy work.
9. B. the Tuskegee Institute in Alabama
10. B. nitrogen
11. cotton
12. A. nitrogen, fixing bacteria grew along their roots
13. false, his dog
14. B. loop

Chapter 8 answers

1. c. insects
   e. spiders and ticks
   b. crabs and lobsters
   a. centipedes
   d. millipedes
2. B. joint
3. A. cricket
4. true
5. false, he observed their behavior in nature
6. head, thorax, abdomen
7. B. mate and reproduce
8. A. butterfly
9. false, Louis Pasteur
10. true
11. A. aphid

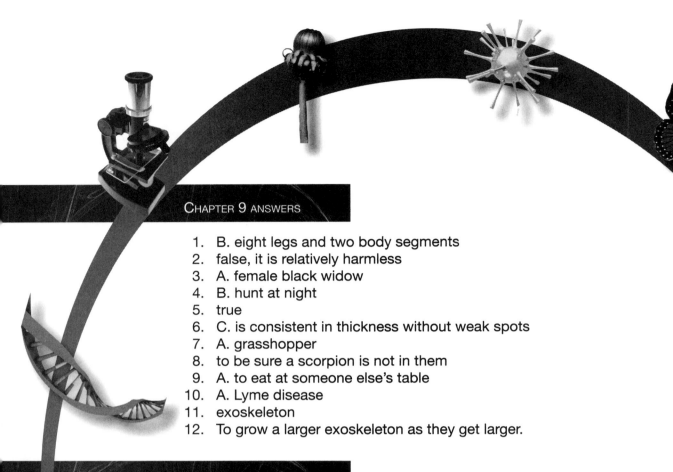

Chapter 9 Answers

1. B. eight legs and two body segments
2. false, it is relatively harmless
3. A. female black widow
4. B. hunt at night
5. true
6. C. is consistent in thickness without weak spots
7. A. grasshopper
8. to be sure a scorpion is not in them
9. A. to eat at someone else's table
10. A. Lyme disease
11. exoskeleton
12. To grow a larger exoskeleton as they get larger.

Chapter 10 Answers

1. false, they continue to be discovered
2. A. invertebrates
3. B. fish
4. D. subphylum
5. A. cold-blooded
6. lateral line
7. B. *The Silent World*
8. C. salamanders
9. skin
10. false, amphibian skin has no scales
11. A. cold-blooded
12. B. decline

Chapter 11 Answers

1. true
2. A. cold-blooded
3. A. frog

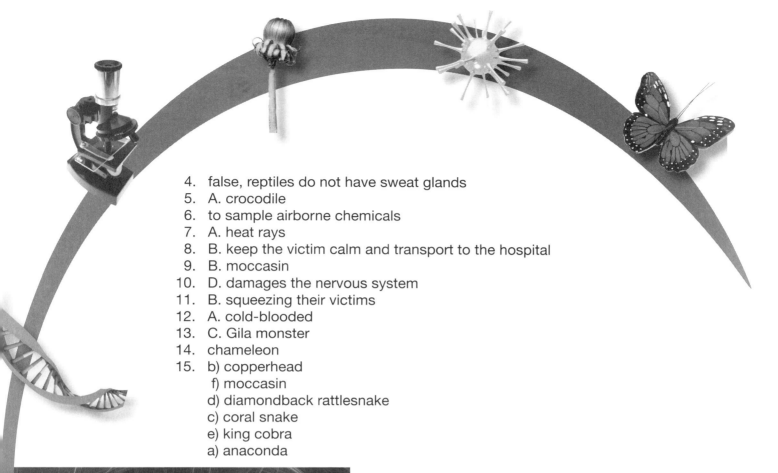

4. false, reptiles do not have sweat glands
5. A. crocodile
6. to sample airborne chemicals
7. A. heat rays
8. B. keep the victim calm and transport to the hospital
9. B. moccasin
10. D. damages the nervous system
11. B. squeezing their victims
12. A. cold-blooded
13. C. Gila monster
14. chameleon
15. b) copperhead
    f) moccasin
    d) diamondback rattlesnake
    c) coral snake
    e) king cobra
    a) anaconda

CHAPTER 12 ANSWERS

1. B. warm-blooded
2. B. more
3. B. have feathers
4. poultry
5. pigeons
6. C. Roger Tory Peterson
7. D. all of the above
8. B. male
9. They need high-energy food.
10. D. Its feathers have no oil coating.
11. gizzard
12. true
13. g) parrot
    e) falcon
    d) dodo
    h) passenger pigeon
    f) hummingbird
    a) anhinga
    b) Arctic tern
    c) cassowary

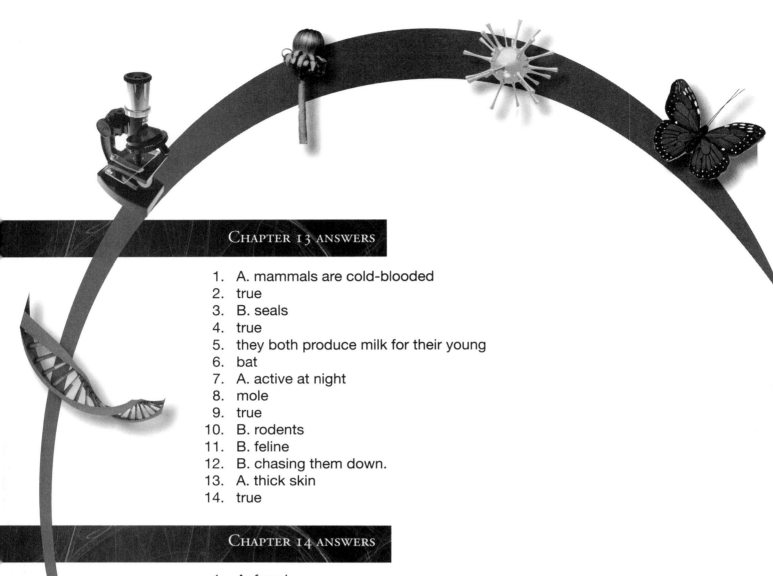

Chapter 13 answers

1. A. mammals are cold-blooded
2. true
3. B. seals
4. true
5. they both produce milk for their young
6. bat
7. A. active at night
8. mole
9. true
10. B. rodents
11. B. feline
12. B. chasing them down.
13. A. thick skin
14. true

Chapter 14 answers

1. A. fraud
2. A. past life
3. C. to make the fossil more valuable
4. A. did not match
5. false
6. true
7. C. part of a skull.
8. A. England
9. true
10. D. wise man
11. False; spear points, a flute, and other items have been found.
12. He had to stoop to go in and never thought to look overhead.

References

Isaac Asimov, *Asimov's BiographicalEncyclopedia of Science and Technology*, Second Revised Edition (Garden City, NY: Doubleday & Company, Inc., 1982).

Michael H. Hart, *The 100: A Ranking of the Most Influential Persons in History* (Secaucus, NJ: Carol Publishing Group, 1993).

Paul de Kruif, *Microbe Hunters* (New York: Harcourt, Brace, and Col., 1926).

A. E. E. McKenzie, *The Major Achievements of Science* (New York: Simon and Schuster, 1960).

Henry M. Morris, *Men of Science, Men of God* (San Diego: Creation-Life Publishers, 1982).

Internet resources:

Caution to parents: The contents of these sites were confirmed at the time of publication. However, Internet sites can change and become inappropriate. Check these sites before allowing your children to view them.

http://www.biography.com/ Biographies of more than 20,000 individuals.

http://nobelprize.org/search/all_laureates_yd.html All about Nobel Prize winners.

http://encarta.msn.com/ An on-line encyclopedia and other resources.

Index

A
abdomen, 76-77, 81, 85, 90, 121, 123
Agassiz, Louis, 66, 73
algae, 21-24, 27, 116
alligators, 100-101
almond, 39-40, 67, 73
amoebas, 19-21
amino acids, 54, 56, 61-65
amphibian, 97-99
anaconda, 107, 109
anaerobic, 24, 27
anhinga, 116, 119
animal kingdom, 6, 13, 21, 35, 37, 75, 87, 92-93, 98, 101, 107, 111, 121, 123-124, 130
annual, 45, 47
antibodies, 54
Aqualung, 96
aquatic, 104
arachnids, 3, 84, 87, 89, 91
Arbor Day, 69
archaeologists, 134-135, 138, 141
archaeoraptor, 132-133
Arctic tern, 117, 119
Aristotle, 32, 37
arthropod, 75, 82-83, 86-88, 90
artificial selection, 67, 73
asexual reproduction, 10
Audubon, John J., 114, 118

B
bacteria, 13-14, 16, 18, 24-27, 51, 54, 60, 63, 71-73, 79, 89, 102,
Bartholdi, Frédéric Auguste, 50
bats, 123-124, 130
Beaumont, William, 59-60
Bell, Alexander Graham, 69
biennial, 45, 47
bird watching, 79, 112-113, 118

Birds of America, 114-115
Black Death, 24, 27
black widow, 85, 90
blubber, 121, 125
brown recluse, 86, 90
Burbank, Luther, 66-67, 69, 72-73
butterflies, 75-80

C
calcium, 121
calorie, 51, 55
camouflage, 90
Canary Islands, 28-30, 36-37
Canine, 30, 57, 124-127, 131, 143
canine teeth, 30, 57, 125-126
Canis Major, 30, 37
carbohydrates, 11, 51-56, 63-64
carbon dioxide, 6, 11, 15, 22, 43, 45, 49, 70, 94, 97
carnivore, 35, 124, 126, 131
carrier pigeon, 111
Carver, George Washington, 69-70, 73
cassowary, 118-119
catalysts, 57
caterpillar, 76-78
cellulose, 6-7, 36, 51, 55
century plant, 40, 47
cereal grains, 48, 54
chameleon, 108-109
Chapman, John, 68, 73
cheetah, 128-130
chiggers, 89
chlorophyll, 6-7, 13, 18, 21-23, 27, 45-46
chordate, 93
cicada, 80, 83
classification, 3, 5-7, 21, 26, 28, 35-36, 75, 77, 87, 92-93, 98, 101, 107, 111, 121, 123-124, 129-130, 136, 141
cobra, 105-106, 109

cocklebur, 39, 71
coconut palm, 41
cocoon, 77-79
coffee, 46-47, 61
cold-blooded, 95, 98, 100, 107-108, 118, 131
collagen, 54, 93, 136
complex organism, 19
copperhead, 103-104, 106, 109
coral snake, 106-107, 109
cottonwood trees, 41
Cousteau, Jacques, 95-97, 99
crocodile, 108, 147
cross-pollination, 67, 69, 73

D
dandelion, 40-41, 47
Dawson, Charles, 135, 137
de Mestral, George, 71, 73
diatoms, 21, 27
digestion, 3, 20, 39, 52, 56-58, 60, 62-65
dodo, 110, 113, 115, 119

E
Edison, Thomas A., 69
elephant, 129-130
embryo, 38, 46
entomologists, 75
enzymes, 7, 20, 54, 57, 62-65, 102
epidermis, 43
esophagus, 58-59, 62, 117
euglena, 21, 23, 27
evergreens, 39
exoskeleton, 77-78, 81, 90

F
Fabre, Jean Henri, 76, 79, 83
falcon, 119, 147
fats, 51-53, 55-56, 63-65
fatty acids, 64-65
fermentation, 9-11
Field Guide to Eastern Birds, 112
fish, 16-17, 32, 37, 53, 63, 92-99, 101, 104-105, 110, 116, 119

flagellum, 21, 23-24
flamingo, 116
Fleming, Alexander, 14-15
Ford, Henry, 69
fraud, 132, 134, 140-141
fructose, 51, 55
fungi, 7-11, 13-15, 22-27, 54

G
gecko, 108-109
germinate, 38
giardiasis, 20-21, 27
Gila monster, 107-109
gills, 8, 94, 97, 99
gizzard, 117
glucose, 45, 52-57, 63-65, 90, 120
glycerol, 64
glycogen, 52-53, 55
grafting, 67, 73
guinea fowl, 31
guinea pig, 30, 33, 37

H
hoax, 132, 135-137, 139-141
Homo sapiens, 138, 141
honey guide, 111
Hooke, Robert, 16-18, 27
hormones, 54, 108
Hugo, Victor, 76
hybridization, 67
hydrochloric acid, 60, 63, 65
hyphae, 7-8, 14

I
insect, 17, 21, 75-78, 81, 83-84, 86-88, 124
insulin, 53
intestine, 63-65, 94
invertebrates, 93, 98

J
Johnny Appleseed, 68

K
keratin, 54
Komodo dragon, 108-109

L
lactose, 51, 55, 65, 120
ladybug, 74, 81-83
lateral line, 94-95, 98
Leeuwenhoek, Anton van, 16, 27
legumes, 25, 71-72
lichen, 15, 22, 27
Linnaeus, Carl, 34-35, 37, 113, 118
lizard, 88, 101, 107-109
locust, 75, 80
Lyme disease, 89, 91

M
maize, 49
malaria, 21, 26, 83
maltose, 51-52, 55
mammal, 120-121, 123, 125, 130-131
matriarch, 129
metamorphosis, 78, 83
Micrographia, 17
milk sugar, 51, 120
mites, 87-89
moccasin, 104, 106, 109
mold, 13-15
moles, 123-124, 130
molting, 90
moth, 74, 78-79
mud puppies, 97
mushrooms, 6-9, 11, 13-16, 27

N
Neanderthal Man, 137
nocturnal, 123, 131
nucleus, 10, 19-20, 23-25, 27
nymph, 78

O
Oakley, Kenneth P., 136
organic, 13, 24-25
ornithologists, 113
oxidation, 49, 51-52, 54

P
pachyderm, 129
paleontologist, 140
pancreas, 53, 63, 94
parakeet, 115
paramecium, 19-20, 27
parasite, 20, 88, 91
parrot, 111, 119
passenger pigeon, 110, 115, 119
Pasteur, Louis, 9-12, 15, 27, 76, 79, 83
pasteurization, 12
penicillin, 13-14
pepsin, 62-63, 65
perennial, 47
Peterson, Roger Tory, 112, 118
Phoenicians, 28-29
phosphorus, 73, 121
photosynthesis, 6, 14, 21, 23, 27, 36, 47
Piltdown man, 135-137, 141
pit viper, 103-104
plague, 24, 26-27
plant kingdom, 6-7, 13-14, 93
Plant Patent Act, 69, 73
popcorn, 49-50
potato, 45, 47, 55, 67, 69, 72
poultry, 111
proboscis, 76
protein, 20, 26, 45, 53-56, 63, 65, 90, 102, 120-121, 127
Protista, 16, 23-24, 26-27
protozoa, 18-21, 23-25, 27
prune, 67, 73
pupa, 77-78
python, 107, 109

Q
quinine, 21

R
rattlesnake, 85, 104-106, 109
reptile, 101-102, 107-108, 122

rhizome, 42
Rocky Mountain spotted fever, 89, 91
rodent, 30, 33, 41, 123, 125
Royal Society, 16, 18, 27, 34, 37
S
salamander, 97
saliva, 56, 64
scorpions, 87-89
SCUBA, 95-96, 98
sheep, 93, 127, 131
Silent World, The, 96, 99
Silk Road, 40, 46
Sirius, 30
Skylab, 87
snakebite, 102
snapdragons, 41
spider, 74-77, 84-88, 90-91
spiny anteater, 121, 131
spiracles, 77
spore, 8-9, 13-14, 42, 47
St. Martin, Alexis, 59-60
staphylococcus, 14
starch, 51-52, 55
stomach, 9, 19, 21, 53, 56, 58-60, 62-65, 94, 117
Strait of Gibraltar, 28, 31, 37
sucrose, 51
sugar, 6, 9-13, 15, 29, 45, 51-57, 61, 65, 90, 120-121
sunflower, 34-35
symbiosis, 22

System of Nature, 34
T
tarantula, 84-85, 90
taste, 7, 9, 13, 15, 43, 51, 56-57, 61, 64-65, 69, 94
teeth, 30, 33, 37, 56-58, 64, 102-103, 117, 119, 125-127, 135-136
termites, 51, 75, 83
thorax, 76-77
tick, 75, 88-89, 91
tiger, 126, 128
toadstools, 9
trilobite, 90
truffles, 7, 9-10, 15
tuberculosis, 24
tumbleweed, 41, 47
Tuskegee Institute, 70, 73
Twenty Thousand Leagues Under the Sea, 95, 99
V
vacuole, 19-20, 23
vegetable, 43-45, 52
vegetative reproduction, 42, 46, 67
Velcro, 71, 73
venom, 84-86, 102-109
Verne, Jules, 95
Virchow, Rudolf, 137
virus, 27
W
warm-blooded, 49, 100, 108, 110, 118, 120-121, 125, 131
Washington, Booker T., 70
whales, 90, 96, 122-125, 130-131
Woodward, Dr. Arthur Smith, 135, 137
X
Xu Xing, 133, 140
Y
yeast, 7, 9-11, 15

From the Center of the Sun to the Edge of God's Universe

Think you know all there is to know about our solar system? You might be surprised!

Master Books is excited to announce the latest masterpiece in the extremely popular *Exploring Series, The World of Astronomy*. Over 150,000 copies of the *Exploring Series* have been sold to date, and this new addition is sure to increase that number significantly!

- Discover how to find constellations like the Royal Family group or those near Orion the Hunter from season to season throughout the year.
- How to use the Sea of Crises as your guidepost for further explorations on the moon's surface
- Investigate deep sky wonders, extra solar planets, and beyond as God's creation comes alive!

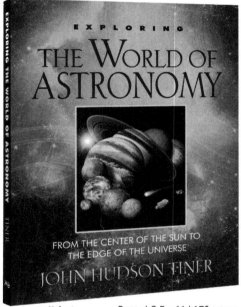

New Edition
Paper | 8.5 x 11 | 176 pages
978-0-89051-787-1 **$14.99**

The book includes discussion ideas, questions, and research opportunities to help expand this great resource on observational astronomy.
Order your copy today to begin *Exploring the World of Astronomy!*
nlpg.com/worldofastronomy

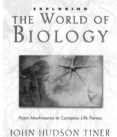

The World of Biology
978-0-89051-552-5
$13.99

The World of Chemistry
978-0-89051-295-1
$13.99

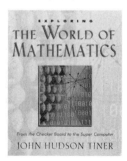

The World of Mathematics
978-0-89051-412-2
$13.99

The History of Medicine
978-0-89051-248-7
$13.99

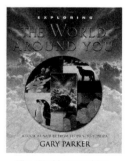

The World Around You
978-0-89051-377-4
$13.99

Planet Earth
978-0-89051-178-7
$13.99

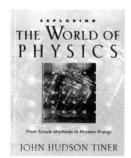

The World of Physics
978-0-89051-466-5
$13.99

nlpg.com/scholarship

Our Award-Winning
Wonders of Creation Series

Filled with special features, every exciting title includes over 200 beautiful full-color photos and illustrations, practical hands-on learning experiments, charts, graphs, glossary, and index — it's no wonder these books have become one of our most requested series.

- **The Ecology Book*** researches the relationship between living organisms and our place in God's wondrous creation.
- **The Archaeology Book*** uncovers ancient history from alphabets to ziggurats.
- **The Cave Book** digs deep into the hidden wonders beneath the surface.
- **The Astronomy Book** soars through the solar system separating myth from fact.
- **The Geology Book** provides a tour of the earth's crust pointing out the beauty and the scientific evidences for creation.
- **The Fossil Book** explains everything about fossils while also demonstrating the shortcomings of the evolutionary theory.
- **The Ocean Book** explores the depths of the ocean to find the mysteries of the deep.
- **The Weather Book** delves into all weather phenomena, including historical weather events.

8 1/2 x 11 • Casebound • 96 pages • Full-color interior
ISBN-13: 978-0-89051-701-7
JR. HIGH to HIGH SCHOOL

*This title is color-coded with three educational levels in mind: 5th to 6th grades, 7th to 8th grades, and 9th through 11th grades and is $16.99. All other titles are available for $15.99.

sample interior from The Archaeology Book

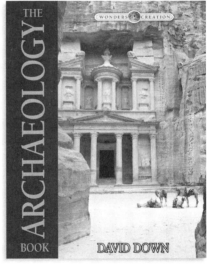

The Archaeology Book
ISBN-13: 978-0-89051-573-0

The Ocean Book
ISBN-13: 978-0-89051-401-6

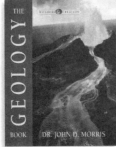

The Geology Book
ISBN-13: 978-0-89051-281-4

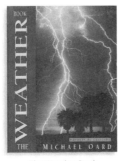

The Weather Book
ISBN-13: 978-0-89051-211-1

The Astronomy Book
ISBN-13: 978-0-89051-250-0

The Fossil Book
ISBN-13: 978-0-89051-438-2

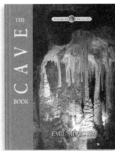

The Cave Book
ISBN-13: 978-0-89051-496-2

Parent Lesson Plan — Promotion

Now turn your favorite **Master Books** into curriculum! Each Parent Lesson Plan (PLP) includes:

- An easy-to-follow, one-year educational calendar
- Helpful worksheets, quizzes, tests, and answer keys
- Additional teaching helps and insights
- Complete with all you need to quickly and easily begin your education program today!

ELEMENTARY ZOOLOGY

1 year
4th – 6th

Package Includes: *World of Animals, Dinosaur Activity Book, The Complete Aquarium Adventure, The Complete Zoo Adventure, Parent Lesson Plan*

5 Book Package
978-0-89051-747-5 $84.99

SCIENCE STARTERS: ELEMENTARY PHYSICAL & EARTH SCIENCE

1 year
3rd – 8th grade

6 Book Package Includes: *Forces & Motion –Student, Student Journal, and Teacher; The Earth – Student, Teacher & Student Journal; Parent Lesson Plan*

6 Book Package
978-0-89051-748-2 $51.99

SCIENCE STARTERS: ELEMENTARY CHEMISTRY & PHYSICS

1 year
3rd – 8th grade

Package Includes: *Matter – Student, Student Journal, and Teacher; Energy – Student, Teacher, & Student Journal; Parent Lesson Plan*

7 Book Package
978-0-89051-749-9 $54.99

INTRO TO METEOROLOGY & ASTRONOMY

1 year
7th – 9th grade
½ Credit

Package Includes: *The Weather Book; The Astronomy Book; Parent Lesson Plan*

3 Book Package
978-0-89051-753-6 $44.99

INTRO TO OCEANOGRAPHY & ECOLOGY

1 year
7th – 9th grade
½ Credit

Package Includes: *The Ocean Book; The Ecology Book; Parent Lesson Plan*

3 Book Package
978-0-89051-754-3 $45.99

INTRO TO SPELEOLOGY & PALEONTOLOGY

1 year
7th – 9th grade
½ Credit

Package Includes: *The Cave Book; The Fossil Book; Parent Lesson Plan*

3 Book Package
978-0-89051-752-9 $44.99

CONCEPTS OF MEDICINE & BIOLOGY

1 year
7th – 9th grade
½ Credit

Package Includes: *Exploring the History of Medicine; Exploring the World of Biology; Parent Lesson Plan*

3 Book Package
978-0-89051-756-7 $40.99

CONCEPTS OF MATHEMATICS & PHYSICS

1 year
7th – 9th grade
½ Credit

Package Includes: *Exploring the World of Mathematics; Exploring the World of Physics; Parent Lesson Plan*

3 Book Package
978-0-89051-757-4 $40.99

CONCEPTS OF EARTH SCIENCE & CHEMISTRY

1 year
7th – 9th grade
½ Credit

Package Includes: *Exploring Planet Earth; Exploring the World of Chemistry; Parent Lesson Plan*

3 Book Package
978-0-89051-755-0 $40.99

THE SCIENCE OF LIFE: BIOLOGY

1 year
8th – 9th grade
½ Credit

Package Includes: *Building Blocks in Science; Building Blocks in Life Science; Parent Lesson Plan*

3 Book Package
978-0-89051-758-1 $44.99

BASIC PRE-MED

1 year
8th – 9th grade
½ Credit

Package Includes: *The Genesis of Germs; The Building Blocks in Life Science; Parent Lesson Plan*

3 Book Package
978-0-89051-759-8 $43.99

800.999.3777

Parent Lesson Plan — Promotion

INTRO TO ASTRONOMY

1 year
7th – 9th grade
½ Credit

Package Includes: *The Stargazer's Guide to the Night Sky; Parent Lesson Plan*

2 Book Package
978-0-89051-760-4 $47.99

INTRO TO ARCHAEOLOGY & GEOLOGY

1 year
7th – 9th
½ Credit

Package Includes: *The Archaeology Book; The Geology Book; Parent Lesson Plan*

3 Book Package
978-0-89051-751-2 $45.99

SURVEY OF SCIENCE HISTORY & CONCEPTS

1 year
10th – 12th grade
1 Credit

Package Includes: *The World of Mathematics; The World of Physics; The World of Biology; The World of Chemistry; Parent Lesson Plan*

5 Book Package
978-0-89051-764-2 $72.99

SURVEY OF SCIENCE SPECIALTIES

1 year
10th – 12th grade
1 Credit

Package Includes: *The Cave Book; The Fossil Book; The Geology Book; The Archaeology Book; Parent Lesson Plan*

5 Book Package
978-0-89051-765-9 $81.99

SURVEY OF ASTRONOMY

1 year
10th – 12th grade
1 Credit

Package Includes: *The Stargazers Guide to the Night Sky; Our Created Moon; Taking Back Astronomy; Our Created Moon DVD; Created Cosmos DVD; Parent Lesson Plan*

4 Book, 2 DVD Package
978-0-89051-766-6 $112.99

GEOLOGY & BIBLICAL HISTORY

1 year
8th – 9th
1 Credit

Package Includes: *Explore the Grand Canyon; Explore Yellowstone; Explore Yosemite & Zion National Parks; Your Guide to the Grand Canyon; Your Guide to Yellowstone; Your Guide to Zion & Bryce Canyon National Parks; Parent Lesson Plan.*

4 Book, 3 DVD Package
978-0-89051-750-5 $112.99

PALEONTOLOGY: LIVING FOSSILS

1 year
10th – 12th grade
½ Credit

Package Includes: *Living Fossils, Living Fossils Teacher Guide, Living Fossils DVD; Parent Lesson Plan*

3 Book, 1 DVD Package
978-0-89051-763-5 $66.99

LIFE SCIENCE ORIGINS & SCIENTIFIC THEORY

1 year
10th – 12th grade
1 Credit

Package Includes: *Evolution: the Grand Experiment, Teacher Guide, DVD; Living Fossils, Teacher Guide, DVD; Parent Lesson Plan*

5 Book, 2 DVD Package
978-0-89051-761-1 $139.99

NATURAL SCIENCE THE STORY OF ORIGINS

1 year
10th – 12th grade
½ Credit

Package Includes: *Evolution: the Grand Experiment; Evolution: the Grand Experiment Teacher's Guide, Evolution: the Grand Experiment DVD; Parent Lesson Plan*

3 Book, 1 DVD Package
978-0-89051-762-8 $66.99

ADVANCED PRE-MED STUDIES

1 year
10th – 12th grade
1 Credit

Package Includes: *Building Blocks in Life Science; The Genesis of Germs; Body by Design; Exploring the History of Medicine; Parent Lesson Plan*

5 Book Package
978-0-89051-767-3 $76.99

BIBLICAL ARCHAEOLOGY

1 year
10th – 12th grade
1 Credit

Package Includes: *Unwrapping the Pharaohs; Unveiling the Kings of Israel; The Archaeology Book; Parent Lesson Plan.*

4 Book Package
978-0-89051-768-0 $99.99

CHRISTIAN HERITAGE

1 year
10th – 12th grade
1 Credit

Package Includes: *For You They Signed; Lesson Parent Plan*

2 Book Package
978-0-89051-769-7 $50.99

Social Media
Win Books & Discover New Resources

facebook.com/**masterbooks**
twitter.com/**masterbooks4u**

nlpgblogs.com

Creation Conversations
Connect with other students who believe in Biblical Creation

creationconversations.com

$3,000 Scholarship

Register for the Master Books Scholarship Essay Program.

Master Books®, a division of New Leaf Publishing Group, will award one $3000 scholarship each year. The Master Books Scholarship Essay Contest is open to any high school junior or senior or the equivalent thereof from any public, private, or homeschool venue. The applicant must be a U.S. citizen and have a minimum GPA of 3.0 or above. This scholarship is a one-time award and may be used at any accredited two-year, four-year, or trade school within the contiguous United States. This award covers only tuition and university-provided room and board.

Learn more at **nlpg.com/scholarship**